U0920545

中国海洋统计年鉴

CHINA MARINE STATISTICAL YEARBOOK 2014

国 家 海 洋 局

Edited by

State Oceanic Administration,

People's Republic of China

海洋出版社

China Ocean Press

《中国海洋统计年鉴》编委会

周　忠　　中国船舶工业集团公司
刘　悦　　中国船舶重工集团公司
姜力孚　　中国石油天然气集团公司
刘　岩　　中国石油化工集团公司
金晓剑　　中国海洋石油总公司
范　志　　中国盐业总公司
乔　梁　　辽宁省海洋与渔业厅
于忠信　　河北省国土资源厅
孙连友　　天津市海洋局
王伟杰　　山东省海洋与渔业厅
张建军　　江苏省海洋与渔业局
沈依云　　上海市海洋局
童加朝　　浙江省海洋与渔业局
陈思增　　福建省海洋与渔业厅
洪伟东　　广东省海洋与渔业局
刘　斌　　广西壮族自治区海洋局
潘建纲　　海南省海洋与渔业厅
栾玉瑄　　大连市海洋与渔业局
傅道军　　青岛市海洋与渔业局
吴建义　　宁波市海洋与渔业局
王春生　　厦门市海洋与渔业局
梁俊乾　　深圳市海洋局

《中国海洋统计年鉴》编辑部

主　任：

何广顺　　国家海洋信息中心副主任

副主任：

沈　君　　国家海洋局战略规划与经济司副司长

王晓惠　　国家海洋信息中心海洋经济部主任

成　员：

崔晓健　王占坤　李长如　杨　娜　郭　越　蔡大浩

周洪军　郑　莉　李巧稚　赵心宇　张　伟　林香红

宋维玲　张潇娴　朱　凌　李琳琳　董　伟　徐丛春

李宜良（国家海洋信息中心）

曹　斐　朱宁东（国家海洋局战略规划与经济司）

特邀编辑：

李燕丽　教育部

林　涛　科技部

苗　强　国土资源部

陈　钟　交通运输部

张淑娜　水利部

郭　睿　农业部

毛玉如　环境保护部

刘建杰　国家林业局

周　鲲　国家旅游局

崔胜先　中国科学院

张淑丽　中国地震局

骆　婷　中国气象局

Editorial Board of
China Marine Statistical Yearbook

Shi Bing	Chinese Academy of Sciences
Han Zhiqiang	China Earthquake Administration
Xie Pu	China Meteorological Administration
Zhou Zhong	China State Shipbuilding Corporation
Liu Yue	China Shipbuilding Industry Corporation
Jiang Lifu	China National Petroleum Corporation
Liu Yan	China Petrochemical Group Corporation
Jin Xiaojian	China National Offshore Oil Corporation
Fan Zhi	China National Salt Industry Corporation
Qiao Liang	Department of Ocean and Fisheries of Liaoning Province
Yu Zhongxin	Hebei Province Department of Land and Resources
Sun Lianyou	Tianjin Ocean Administration
Wang Weijie	Department of Oceanic and Fishery of Shandong Province
Zhang Jianjun	Jiangsu Provincial Ocean and Fisheries Bureau
Shen Yiyun	Shanghai Ocean Administration
Tong Jiachao	Zhejiang Provincial Ocean and Fisheries Bureau
Chen Sizeng	Department of Oceanic and Fishery of Fujian Province
Hong Weidong	Guangdong Provincial Oceanic and Fishery Administration
Liu Bin	The Oceanic Administration of Guangxi
Pan Jiangang	Department of Marine and Fishery of Hainan Province
Luan Yuxuan	Dalian Ocean and Fishery Administration
Fu Daojun	Qingdao Ocean and Fishery Administration
Wu Jianyi	Ningbo Ocean and Fishery Bureau
Wang Chunsheng	Oceans and Fisheries Bureau of Xiamen
Liang Junqian	Shenzhen Ocean Administration

Editorial Department of China Marine Statistical Yearbook

Qiu Yinlang	China State Shipbuilding Corporation
Xu Jing	China Shipbuilding Industry Corporation
Sun Xiaoxian	China National Petroleum Corporation
Sun Yan	China Petrochemical Group Corporation
Hao Jinghui	China National Offshore Oil Corporation
Du Lihua	China National Salt Industry Corporation
Wei Nan	Department of Ocean and Fisheries of Liaoning Province
Cheng Xin	Hebei Province Department of Land and Resources
Yang Jinlu	Tianjin Ocean Administration
Liu Haibin	Department of Oceanic and Fishery of Shandong Province
Pei Pei	Jiangsu Provincial Ocean and Fisheries Bureau
Zhang Cheng	Shanghai Municipal Maritime Affairs Administration
Wang Zhiyong	Zhejiang Provincial Ocean and Fisheries Bureau
Liao Runxue	Department of Oceanic and Fishery of Fujian Province
Li Hongbing	Guangdong Provincial Oceanic and Fishery Administration
Cao Shuping	The Oceanic Administration of Guangxi
Lin Yun	Department of Marine and Fishery of Hainan Province
Shen Ronglian	Dalian Ocean and Fishery Administration
Xiao Bentong	Qingdao Ocean and Fishery Administration
Wu Mingchun	Ningbo Ocean and Fishery Bureau
Chi Xincai	Oceans and Fisheries Bureau of Xiamen
Luo Xiaoxia	Shenzhen Municipal Economic, Trade and Informanization Commission

Executive Editor: Li Changru

English Proof-reader: Lin Baofa

编 者 说 明

一、《中国海洋统计年鉴》2014 年版是一部全面反映 2013 年中华人民共和国海洋经济发展和海洋管理服务情况的资料性年鉴，全书为中英文对照。

二、本年鉴的统计资料范围为人们在海洋和沿海地区开发、管理、利用海洋资源和空间，发展海洋经济的生产和活动以及沿海地区的社会经济概况。地域范围为沿海地区、沿海城市和沿海地带，其排列顺序按《沿海行政区域分类与代码》(HY/T 094-2006) 的顺序排列。

三、本年鉴内容包括综合资料、海洋经济核算、主要海洋产业活动、主要海洋产业生产能力、涉海就业、海洋科学技术、海洋教育、海洋环境保护、海洋行政管理及公益服务、全国及沿海社会经济、部分世界海洋经济统计资料等十一部分。

四、本年鉴根据《海洋统计报表制度》（国统制 [2012] 50 号）和《海洋生产总值核算制度》（国统制 [2013] 93 号），资料主要来源于沿海省、自治区、直辖市统计局、海洋厅（局）以及 20 个有关涉海部、局、总公司。

五、本年鉴中除特殊说明外，所有价值量指标均为当年价，除注明年份外，其他均为 2013 年数据，年鉴中每部分后附有主要海洋统计指标解释，对主要海洋统计指标的含义、统计范围和统计方法做了简要说明。统计数据中的其他说明置于表的下面。有续表的资料，如有注释均置于最后一张表的下面。

六、由于全国第三次经济普查数据尚在审核汇总中，故本年鉴中“全国及沿海社会经济”和“海洋经济核算”中有关数据为初步核算数。

七、本年鉴表格中符号使用说明：“…”表示数据不足本表最小计算单位数；

“空格”表示该项统计指标数据不详或无该项数据；“#”表示其中的主要项；其他符号如“*”或“①”等表示本表后面有注释。

八、本年鉴中由于数字精确度的原因，四舍五入后部分分项值之和与合计值有微小差异。

九、本年鉴资料国内部分未包括香港特别行政区、澳门特别行政区和台湾省数据。

十、《中国海洋统计年鉴》在编撰过程中，得到了各有关单位的大力支持，我们在此表示衷心的感谢。本年鉴中如有疏漏和不妥之处，敬请读者批评指正。

《中国海洋统计年鉴》编辑部

Introduction

I. *China Marine Statistical Yearbook* (2014) is a data almanac which reflects in an all-round way the development of marine economy, marine management and service in the People's Republic of China in 2013, and it is a Chinese-English bilingual edition.

II. The Yearbook's statistics cover the production and activities in the marine and coastal areas in relation to the development, management and utilization of marine resources and space, and the development of marine socioeconomy. The regions covered are the coastal regions, coastal cities and coastal zones with coastlines, which are arranged in order according to the *Coastal Administrative Areas Classification and Codes* (HY/T 094—2006).

III. The data in the yearbook consist of 11 sections, namely, integrated data, marine economic accounting, major marine industrial activities, production capacity of major marine industries, ocean-related employment, marine science and technology, marine education, marine environmental protection, marine administration and public-good service, national and coastal socioeconomy, part of the world's marine economic statistics data.

IV. The Yearbook is based on the *Marine Statistics Report System* (Guotongzhi [2012] No. 50) and the *Ocean Gross Product Accounting System* (Guotongzhi [2013] No. 93), its data mainly come from the statistical bureaus and the oceanic administrations of the coastal provinces, autonomous regions, and municipalities directly under the Central Government as well as the 20 ocean-related ministries, bureaus and general corporations concerned.

V. Unless otherwise specified in the Yearbook, all the value indicators are given at the current price. Except for those noted in years, all the others are the data of 2013. Each section is attached by explanatory notes to the major marine statistical indicators, giving a brief explanation for the meaning, statistical range and statistical methods of the major marine statistical indicators. Other notes to the statistical data are listed below the tables. For the data with continued tables, annotations, if any, are put below the last table.

VI. Since the data of the Third National Economic Census are still under review and accumulation, the relevant data in the chapters of National and Coastal Socioeconomy and Marine Economic Accounting are the preliminary accounting numbers.

VII. The usage of symbols in the tables: "…" indicates the statistics smaller than the minimum calculation unit in the table; "Blank" indicates that the data of the statistical index is unknown for the time being or that there is not such data available; "#" indicates the major items of the table; Other symbols, such as "*" or "①", indicate "see footnotes below".

VIII. For reasons of digital accuracy, there is small difference between the sum of values of some subterms after having been rounded off and the total values.

IX. The domestic part of the Yearbook does not include that of Hong Kong Special Administrative Region, Macau Special Administrative Region and Taiwan Province.

X. In the course of editing *China Marine Statistical Yearbook*, we enjoyed energetic support from the various departments concerned and we hereby extend our heartfelt thanks to them. Criticisms and comments are welcome from readers on any of the oversights and inappropriateness as the time for editing is too short.

Editorial Department of
China Marine Statistical Yearbook

目　次
CONTENTS

3 主要海洋产业活动
Major Marine Industrial Activities

5 涉海就业
Ocean-Related Employment

6 海洋科学技术
Marine Science and Technology

7 海洋教育

Marine Education

9 海洋行政管理及公益服务
Marine Administration and Public-Good Service

10 全国及沿海社会经济
National and Coastal Socioeconomy

11 部分世界海洋经济统计资料

Part of the World's Marine Economic Statistics Data

2013 年我国海洋经济发展综述

2013 年，各级海洋行政主管部门深入贯彻落实党的十八大和十八届三中全会精神，紧紧围绕建设海洋强国的战略部署，加快推进海洋经济发展方式转变，海洋经济继续保持良好发展势头。

一、全国海洋经济发展概况

2013年，全国海洋生产总值54 313亿元，比2012年增长7.6%（除特别注明外，增长率均按可比价计算），海洋生产总值占国内生产总值的9.5%，占沿海地区生产总值的15.8%。全国涉海就业人员3 514万人，比2012年增加45.5万人。

二、主要海洋产业发展情况

2013年，主要海洋产业实现增加值22 681亿元，比2012年增长6.7%，占海洋生产总值的41.8%，滨海旅游业和海洋交通运输业仍占主导地位。

海洋第一产业　2013年，海洋渔业平稳发展，海洋水产品产量稳步增长，海水养殖及远洋渔业生产能力不断提高。海水产品产量为3 138.8万吨，比2012年增长3.5%。其中，海水养殖产量继续增加，达到1 739.2万吨，比2012年增长5.8%；而海洋捕捞产量略有下降，为1 264.4万吨，比2012年降低0.2%；远洋捕捞产量明显提高，达135.2万吨，比2012年增长10.5%。海水养殖面积达到231.56万公顷，比2012年增加6.2%。远洋渔船数量达到2 159艘，比2012年增加20.4%；总功率达到158.35万千瓦，比2012年增长42.4%。

海洋第二产业　2013年，海洋油气业保持稳定发展，海洋原油产量

4 541.1万吨，比2012年增长2.2%，海洋天然气产量117.6亿立方米，比2012年减少4.2%。全年实现增加值1 648亿元，比2012年增长0.1%。海洋矿业较快发展，海洋矿产资源开采秩序进一步规范，全年实现增加值49亿元，比2012年增长13.7%。海洋盐业呈现负增长，全年实现增加值56亿元，比2012年减少8.1%。海洋化工产业运行平稳，全年实现增加值908亿元，比2012年增长11.4%。海洋生物医药产业持续较快发展，全年实现增加值224亿元，比2012年增长20.7%。海洋电力业稳步发展，海上风电项目有序推进，全年实现增加值87亿元，比2012年增长11.9%。海水利用业较快发展，产业技术应用和推广不断加快，产业化水平进一步提高，全年实现增加值12亿元，比2012年增长9.9%。海洋船舶工业生产经营形势依然严峻，经济效益持续下滑，全年实现增加值1 183亿元，比2012年减少7.7%。海洋工程建筑业继续保持稳步增长，全年实现增加值1 680亿元，比2012年增长9.4%。

海洋第三产业　2013年，全国沿海港口生产形势总体良好，但受全球经济增长放缓、国际海运市场低迷拖累以及运力过剩严重等因素影响，我国航运业复苏仍需时日。全年海洋交通运输业实现增加值5 111亿元，比2012年增长4.6%。沿海港口货物吞吐量756 129万吨，比2012年增长9.9%，国际标准集装箱吞吐量16 968万标准箱，比2012年增长7.4%。滨海旅游产业规模继续扩大，沿海地区在加大旅游基础设施投入的同时，不断丰富海洋旅游产品，海洋邮轮游艇旅游发展步入新阶段。但沿海地区入境旅游形势不容乐观，主要沿海城市接待入境旅游者人数3 779.7万人次，比2012年减少769.3万人次；主要沿海城市国际旅游（外汇）收入293.1亿美元，仅比2012年增长0.8%。滨海旅游业通过不断深挖国内旅游市场，全年实现增加值7 851.4亿元，比2012年增长11.7%。

三、区域海洋经济发展情况

2013年，环渤海、长三角、珠三角三大经济区海洋经济继续保持平稳增长的态势，但受世界经济持续低迷和国内经济增速放缓的影响，大部分沿海地区海洋经济增速出现不同程度的回落。环渤海经济区、长江三角洲经济区和珠江三角洲经济区海洋生产总值分别为19 734.0亿元、16 484.8亿元和11 283.6亿元，占全国海洋生产总值的比重分别为36.3%、30.4%、20.8%。

环渤海经济区海洋经济总体增长平稳，海洋生产总值比2012年增长10.1%（现价），占地区生产总值比重15.9%。海洋产业增加值为11 495.3亿元，海洋相关产业增加值为8 238.7亿元。海洋渔业、海洋油气业、海洋交通运输业、滨海旅游业依然是该区海洋经济的支柱产业，增加值合计达到7 631.7亿元，占该地区主要海洋产业增加值的83.2%。海洋生物医药业、海水利用业、海洋电力业实现快速增长，分别与上年增长28.0%、19.5%和12.5%（现价）。

长江三角洲经济区增速继续放缓，海洋生产总值比2012年增长5.6%（现价），海洋生产总值占地区生产总值比重13.9%。长江三角洲海洋产业增加值9 604.2亿元，海洋相关产业增加值6 880.7亿元。按照产值贡献，滨海旅游业、海洋交通运输业、海洋船舶工业和海洋渔业四个产业位居前列，其增加值之和占该地区主要海洋产业增加值的91.2%，其中滨海旅游业占该地区主要海洋产业增加值的42.5%，产值贡献位居第一。海洋电力业、海洋生物医药业、海洋矿业增速较快，分别比2012年增长11.6%、11.4%和9.7%（现价）。

2013年，珠江三角洲经济区着力优化海洋开发布局，构建现代海洋产业体系，建设海洋生态文明，推动增长动力转化和发展方式转变，促进了海洋经济平稳运行，在地区经济和全国海洋经济发展中发挥着重要作用。2013年，珠江三角洲海洋生产总值达11 283.6亿元，比2012年增长7.4%（现价），海洋生产总值占地区生产总值比重达18.2%。海洋产业增加值为6 852.3亿元，海洋相关产业增加值为4 431.4亿元。滨海旅游业、海洋交通运输业、海洋化工业、海洋油气业和海洋渔业依然为珠江三角洲经济区海洋经济发展的支柱产业，其增加值之和占该地区主要海洋产业增加值90.7%。

四、海洋科研教育

2013年，海洋科研教育事业继续保持稳步发展，统计的海洋科研机构共175个，从业人员38 754人，比2012年增长2.9%；海洋科研机构承担海洋科技课题16 331项，比2012年增长6.0%；发表海洋科技论文16 284篇，出版海洋科技著作384种，其中海洋科技著作比2012年增长13.6%；专利授权数3 430件，其中发明专利2 246件，分别比2012年增长24.9%和19.3%。开设海洋专业的高等院校达393个，比上年增长8.9%，专任教师数299 057人，比上年增长9.9%。高等教育和中等职业教育海洋专业毕业生数94 545人，比上年下降10.9%；招生人数和在校人数分别为74 562人和256 093人，分别比上年下降10.5%和6.0%。

五、海洋环境保护

2013年，我国海洋环境质量状况总体维持在较好水平。符合第一类海水水质标准的海域面积约占我国管辖海域面积的95%，海洋沉积物质量状况总体良好。海水、海洋沉积物、海洋生物的放射性水平和海洋大气γ辐射空气吸收剂量率均处于本底范围内。国家级海洋保护区环境质量总体良好，主要保护对象或保护目标基本保持稳定。重点海水浴场、滨海旅游度假区环境质量总体良好。海水增养殖区环境质量基本满足养殖活动要求。海洋倾倒区环境状况总体稳定，未因倾倒活动产生明显影响，近岸海域平均富营养化状况为轻度富营养。但是，近岸海域海水污染、海洋生境退化、环境灾害多发等问题依然突出，陆源排污压力巨大，近岸局部海域污染严重；77%实施监测的河口、海湾等典型海洋生态系统处于亚健康和不健康状态；赤潮发现次数和累计面积为近5年来最少；渤海滨海平原地区海水入侵和土壤盐渍化严重；局部岸段侵蚀程度加大。2013年，我国海洋灾害以风暴潮、海浪、海冰和赤潮灾害为主，绿潮、海岸侵蚀、海水入侵与土壤盐渍化、咸潮入侵等灾害也均有不同程度发生。各类海洋灾害造成直接经济损失163.48亿元，死亡（含失踪）121人。

六、海洋行政管理

2013年，行政管理工作扎实推进，各项海洋工作取得明显成效。全年颁发海域使用权证书3 937本，比2012年增长62.8%；确权海域面积354 979公顷，比2012年增长25.3%；全年共签发疏浚物海洋倾倒许可证116份，实施各项海洋执法检查共166 588次，发现违法行为2 228起；提供海洋数值预报服务共76 082次，开展海洋调查项目569个，获得数据共232.96万个，各项海洋观测获得数据量共52 279.30万个，全年接收存档卫星遥感数据量共计25 714.28GB；本年接收纸质档案23 952卷（册），电子档案4 089GB；共有21项国家标准和72项行业标准通过立项审查，发布国家标准1项，行业标准41项。

Summary of China's Marine Economic Development in 2013

In 2013, the competent marine administrative departments at all levels thoroughly implement the spirit of the 18th National Congress of the Party and the Third Plenary Session of the 18th Central Committee of the Party, and closely around the strategic plan of building a maritime power, speed up the advancement of the transformation of the pattern of marine economic development so that the marine economy continues to keep a good momentum of development.

I. Survey of the National Marine Economic Development

In 2013, the national gross ocean product is 5431.3 billion yuan (RMB), 7.6% up from that in 2012 (Unless otherwise specified, the growth rate is calculated at the comparable price.), accounting for 9.5% of the GDP and 15.8% of the Gross Regional Product. The number of people employed in the ocean related sectors throughout the country reaches 35.14 million, 0.455% million more than that in 2012.

II. Development of Major Marine Industries

In 2013, the main marine industries have realized an added value of 2 268.1 billion yuan, registering an increase of 6.7% as compared with that in 2012, accounting for 41.8% of the gross ocean product, and the coastal tourism and marine communications and transportation still occupy the leading position.

Primary marine industry

In 2013, marine fishery develops smoothly, and the output of marine aquatic products grows steadily and the production capacity of mariculture and pelagic fishery has improved constantly. The production of marine aquatic products is 31.388 million tons, 3.5% up from that in 2012, among which, the yield from mariculture continues to increase, amounting to 17.392 million tons, 5.8% up from that in 2012; the yield from marine fishing drops slightly, amounting to 12.644 million tons, 0.2% down from that in 2012; the output of pelagic fishing has significantly increased, reaching 1.352 million tons, increasing by 10.5% as against that in 2012. The area of mariculture has reached 2.315 6 million hm^2, 6.2% up from that in 2012. the number of the ocean-going fishing vessels amounts to 2 159, 0.4% up from that in 2012; and the total power 1.583 5 million KW, growing by 42.4% as compared with that in 2012.

Secondary marine industry

In 2013, the offshore oil and gas industry has kept a stable development. The output of marine crude oil amounts to 45.411 million tons, 2.2% up from that in 2012 and the production of offshore natural gas 11.76 billion m^3, 4.2% down from that in 2012. An added value of 164.8 billion yuan is effected for the whole year, growing by 0.1% as against that in 2012. The marine mining industry and the order of exploiting marine mineral resources has been further standardized, effecting an added value of 4.9 billion yuan throughout the year, 13.7% up from that in 2012. Depicting a negative growth, the marine salt industry obtains an added value of 5.6 billion yuan, decreasing by 8.1% as compared with that in 2012. The marine chemical industry is functioning smoothly, affecting a full-year added value of 90.8 billion yuan, registering an increase of 11.4% as against that in 2012. The marine biomedicine industry continues to develop rapidly, accomplishing an added value of 22.4 billion yuan for the whole year, 20.7% up from that in 2012. The marine electric power industry has developed steadily and the offshore wind power projects are arrived forward orderly, realizing an added-value of 8.7 billion yuan throughout the year, 11.9% up from that in 2012. The seawater utilization industry is developing rapidly, the application and popularization of industrial technologies is speeded up constantly and the level of industrialization has been further raised, effecting a full-year added value of 1.2 billion yuan, 9.9% up from that in 2012. The situation of production and management of the marine shipbuilding industry is still for bidding, and the economic benefits continue to decline, accomplishing a full-year added value of 118.3 billion yuan, 7.7% down from that in 2012. The marine engineering construction industry continues to keep a steady growth, effecting a full-year added value of 168.0 billion yuan, 9.4% up from that in 2012.

Tertiary marine industry

In 2013, the production situation of national coastal harbours is on the whole good, but, affected by such factors as the slowdown of the global economic growth, the international shipping market downturn and the seriously excess capacity, etc., the recovery of China's shipping industry still needs time. The marine communications and transportation industry effects a full-year added value of 511.1 billion yuan, 4.6% up from that in 2012. The cargo handling capacity of coastal harbours is 7.561 29 billion tons, 9.9% up from that in 2012. And the handling capacity of international standardized containers 169.68 million standard cases, 7.4% up from that in 2012. The Scale of the coastal tourism industry continues to expand, and while increasing the input into the tourist infrastructure, the coastal regions are constantly enriching the marine tourist products and the tourism of oceanic line steamers and cruise yachts has ushered in a new stage of development. However, the situation of inbound tourism in the coastal regions is not optimistic, and the number of inbound tourists received in the major coastal cities is 37.797 million person-times, 7.693 million people less than that in 2012; The income (foreign exchange) from international tourism in the major coastal cities is 29.31 billion US dollars, 0.8% only up from that in 2012. By means of exploiting the domestic tourist market deeply, the coastal tourism industry effects a full-year added value of 785.14 billion yuan, 11.7% up from that in 2012.

III. Development of Regional Marine Economy

In 2013. marine economy in the three major economic zone of the Round-the-Bohai Economic Zone, Changjiang River Delta Economic zone and Zhujiang River Delta Economic Zone contimes to keep a posture of steadily growth, but affected by the sustained downturn of world economy and the slowdown of the growth rate of domestic economy, the growth rate of marine economy in most coastal regions has declined in varying degrees. The gross ocean products of the Round-the-Bohai Economic Zone, Changjiang River Delta Economic Zone and Zhujiang River Delta Economic Zone are 1 973.4 billion yuan, 1 648.48 billion yuan and 1 128.36 billion yuan respectively, accounting for 36.3%, 30.4% and 20.8% of the Gross Ocean Product separately.

The overall growth of marine economy of the Round-the-bohai Economic Zone is stable and the Gross Ocean Product has increased by 10.1% as against that in 2012 (at the current price), accounting for 15.9% of the Gross Regional Product. The added value of marine industries is 1 149.53 billion yuan and that of the marine related industries is 823.87 billion yuan. Marine fishery, offshore oil and gas industry, marine communications and transportation and coastal tourism are still the pillar industries of the marine economy of this zone and their values added amount to 763.17 billion yuan, accounting for 83.2% of those of the major marine industries in the zone. Marine biomedicine, seawater utilization industry, and marine electric power generation industry have grown rapidly, 28.0%, 19.5% and 12.5% from those in the previous year (at the current price).

The growth rate in the Changjiang River Delta Economic Zone continues to slow down, and its gross ocean product increases by 5.6% (at the current price), occupying 13.9% of the gross regional product. The value added of the marine industries reaches 960.42 billion yuan, and that of the marine related industries 688.07 billion yuan Interims of output value contributions, coastal tourism, marine communications and transportation, marine shipbuilding industry and marine fishery rank among the first, their sum total accounting for 91.2% of the values added of the major marine industries in the zone, among which, that of coastal tourism 42.5%, its output value contribution ranking the first. Marine electric power generation industry, marine biomedicine and marine mining grow faster, 11.6%, 11.4% and 9.7% up from those in 2012 (at the current price).

In 2013, the Zhujiang River Delta Economic Zone focuses on optimizing the distribution of marine development, constructing a modern marine industry system, building marine eco-civilization, and pushing forward the transformation of growth momentum and the change of the pattern of development, thus promoting the steady running of marine economy and playing an important role in the development of regional economy and national marine economy. In 2013, the gross ocean product in the Zhujiang River Delta Economic Zone amounts to 1 128.36 billion yuan, 7.4% up from that in 2012 (at the current price), accounting for 18.2% of the gross regional product. The value added of marine industries is 685.23 billion yuan and that of the marine related industries 443.14 billion yuan. Coastal tourism, marine communications and transportation, marine chemical industry, offshore oil and gas industry and marine fishery remain the pillar industries of marine economic development in the zone and the sum total of their value added occupies 90.7% of those of the major marine industries in the zone.

IV. Marine Scientific Research and Education

In 2013, marine scientific research and education continue to develop steadily. The statistical number marine scientific research institutions totals 175 with 38 754 employees, increasing by 2.9% as against that in 2012; the number of marine scientific and technological projects undertaken by these institutions is 16 331, 6.0% up from that in 2012; 16 284 marine scioutifical and technological paper and 384 kinds of marine scientific and technologic work are published, the latter registering an increase by 13.6% as compared with that in 2012; the number of patents authorized is 3 430, of which 2 246 is the patents for discovery, increasing by 24.9% and 19.3% separately as against those in 012. The number of higher learning institutions which offer marine specialities accounts to 393, 8.9% more than that in 2012 and the number of full-time teachers in this regard is 229 057, increasing by 9.9% as compared with that in 2012. The number of graduates from the marine speciality of higher learning and secondary vocational education is 94 545, 10.9% down from that in 2012; and the number of students enrolled and that in school are 74 562 and 256 093, 10.5% and 6.0% down from the previous year respectively.

V. Marine Environmental Protection

In 2013, China's marine environmental quality is on the whole maintained at a fairly good level. The area of the seawaters that is up to the standard of first-class seawater quality accounts for 95% of the sea area under China's jurisdiction, and the quality of marine sediment is, on aggregate, fine. Both the radioactivity level of seawater, marine sediment and marine life, and the rate of air absorbed dose of the marine atmospheric γ radiation are within the range of background. The environmental quality of the national marine protected areas (MPA) is good on the whole and the main objects or targets of protection basically keep stable. The environmental quality in the key bathing beaches and coastal tourists resorts is good as a whole. The environmental quality of in the seawater culture and stock enhancement zone can basically satisfy the demanel for cultivating activities. The environmental condition of the ocean dumping zone is stable on the whole and has not been significantly affected by dumpling activities. The nearshore sea area is in a state of light eutrophication on the arerage.. However, the environmental problems in the nearshore sea area such as seawater pollution, marine habitat degradation, frequent occurrence of disasters, etc. are still prominent. The pressure of ferruginous sewage discharge is great and the local nearshore waters is seriously polluted; 77% of the typical marine ecosystems that have been monitored such as estuaries, bays etc. is in a sub-healthy or unhealthy state; the frequency of discovering the red tide and the grand total of the area where red tides are found are the lowest in the past five years; in the coastal plain area of the Bohai Sea;

seawater intrusion and soil salinization are serious; and in some coastal sections, the degree of erosive has increased. In 2013, the marine disasters are mainly storm surge, sea wave and red tide, and other disasters have happened in varying degrees such as green tide, coastal erosion, seawater intrusion and soil stalinization, salt water intrusion, the direct economic loss caused by various marine disasters reaches 16.348 billion yuan, with a death toll (including missing) of 121 people.

VI. Marine Administration

In 2013, the administrative management work has made solid headway and all the marine work has achieved obvious achievements. A total of 3 937 certificates for the sea area use right are issued throughout the year, 62.8% up from that in 2012; the sea area with the ownership of patent rights reaches 354 979 hm^2, increasing by 25.3% as against that in 2012; a total of 116 permits for oceanic dumping of dredged material are signed and issued for the whole year, and marine inspections for law enforcement are carried out on 166 588 occasions, during which 2 228 cases of unlawful practice one discovered; the marine numerical forecast service is provided for 76 082 times, 569 marine survey project are carried out, acquiring 2.329 6 million data and the data quantity from various marine observations totals 522.793 0 million and the quantity of satellite remote-sensing data received and placed on file for the whole year totals 25 714.28 GB; the paper achieves received this year amounts to 23 952 volumes (copies) and the electronic archives 4 089 GB; and a total of 21 items of national standard and 72 items of professional standard have been authorized and examined, and one item of national standard and 41 items of professional standard issued.

图 1　全国海洋生产总值及三次产业构成
China's Gross Ocean Product and Three Industries Composition

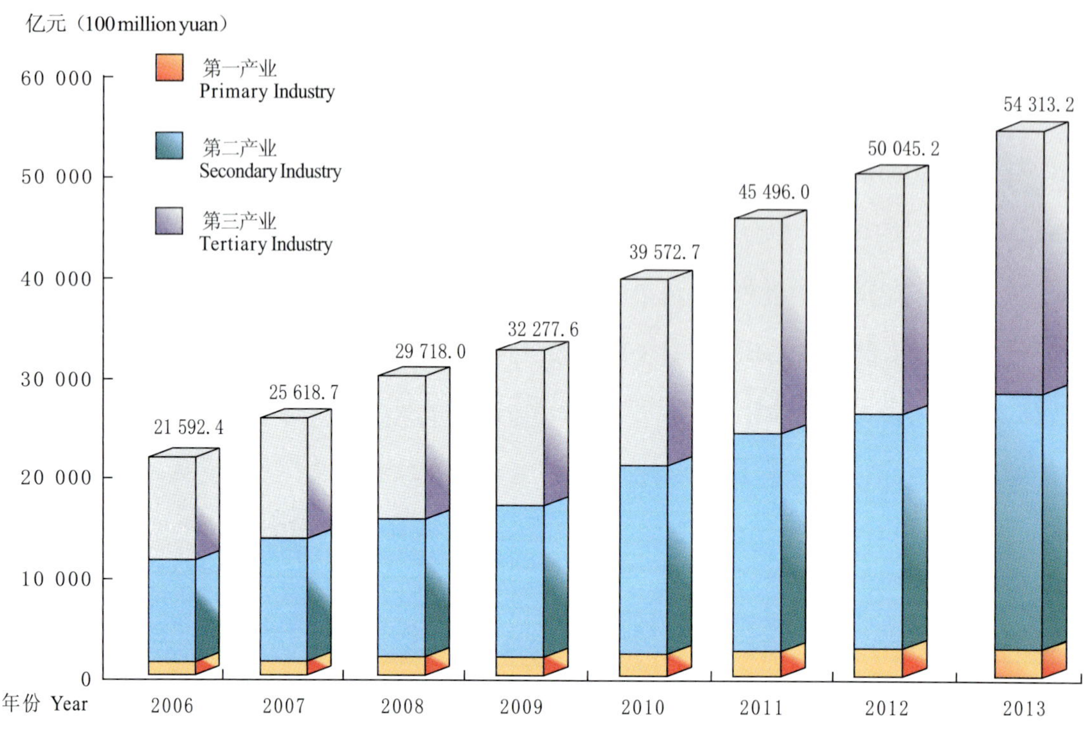

图 2　2013年全国主要海洋产业增加值构成
Composition of Added Values of the Major Marine Industries in China in 2013

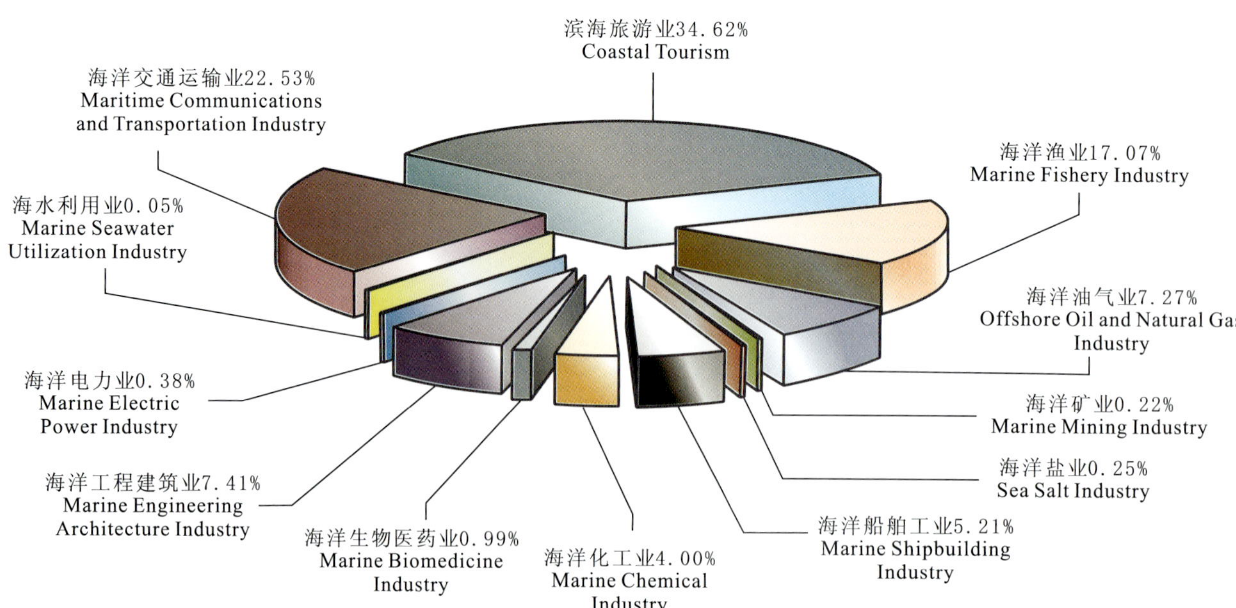

图 3 2013年沿海地区海洋生产总值
Gross Ocean Product by Coastal Regions in 2013

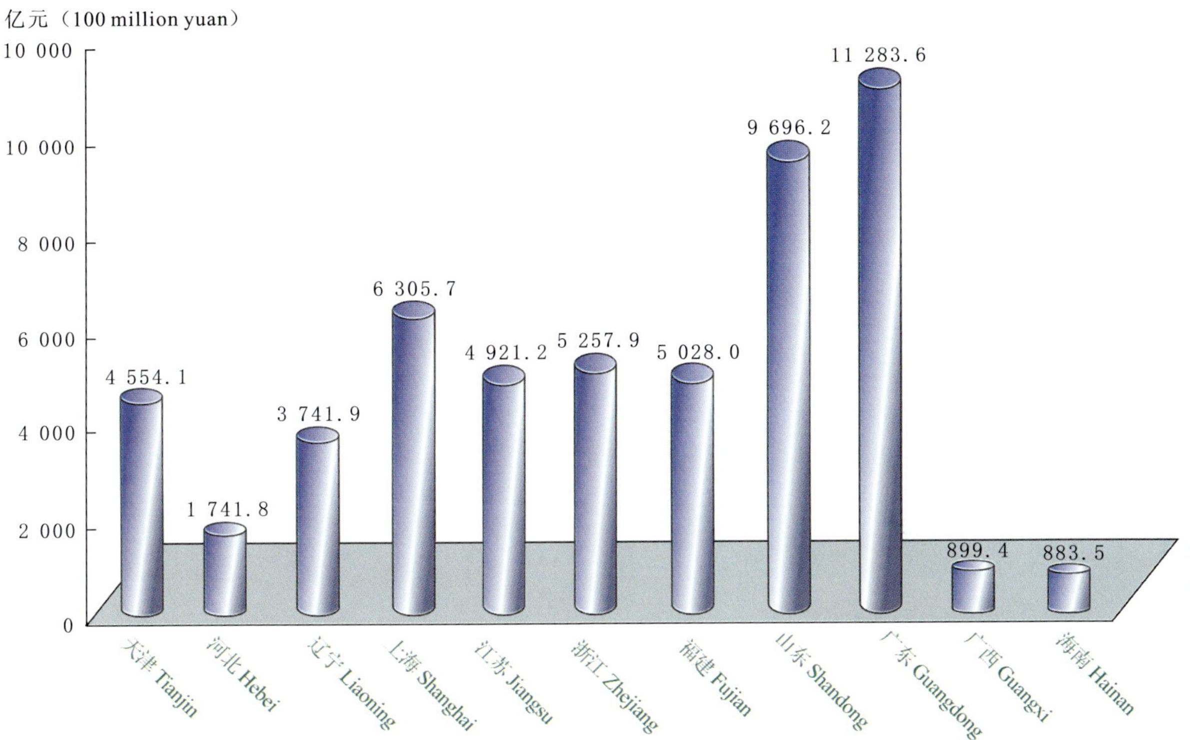

图 4 全国海洋捕捞养殖产量
National Marine Catches and Mariculture Production

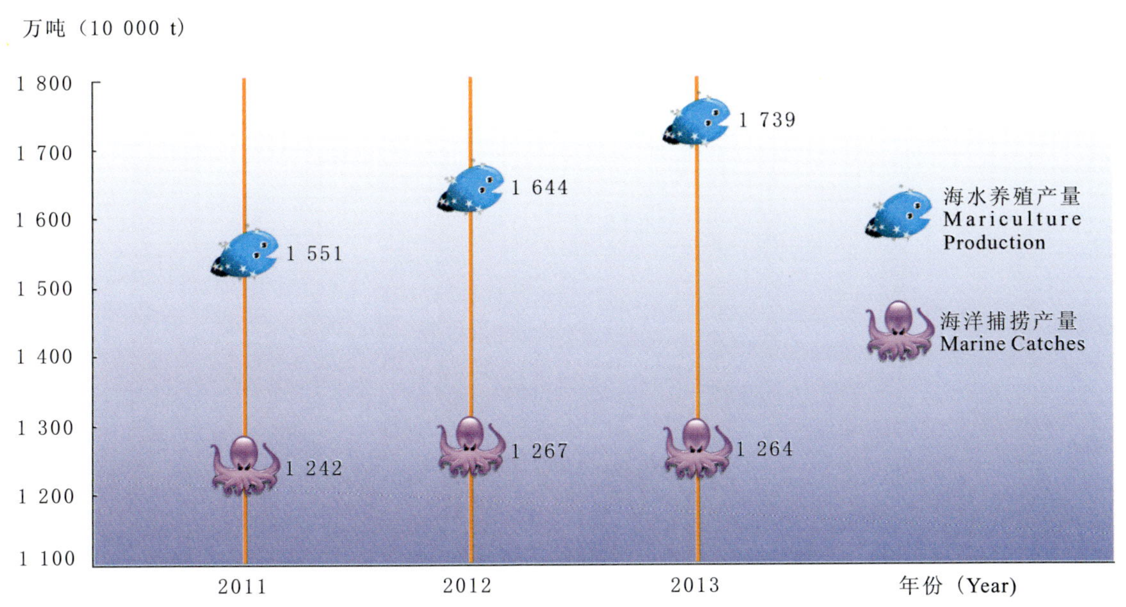

图 5 2013年沿海地区海洋捕捞养殖产量

Marine Catches and Mariculture Production by Coastal Regions in 2013

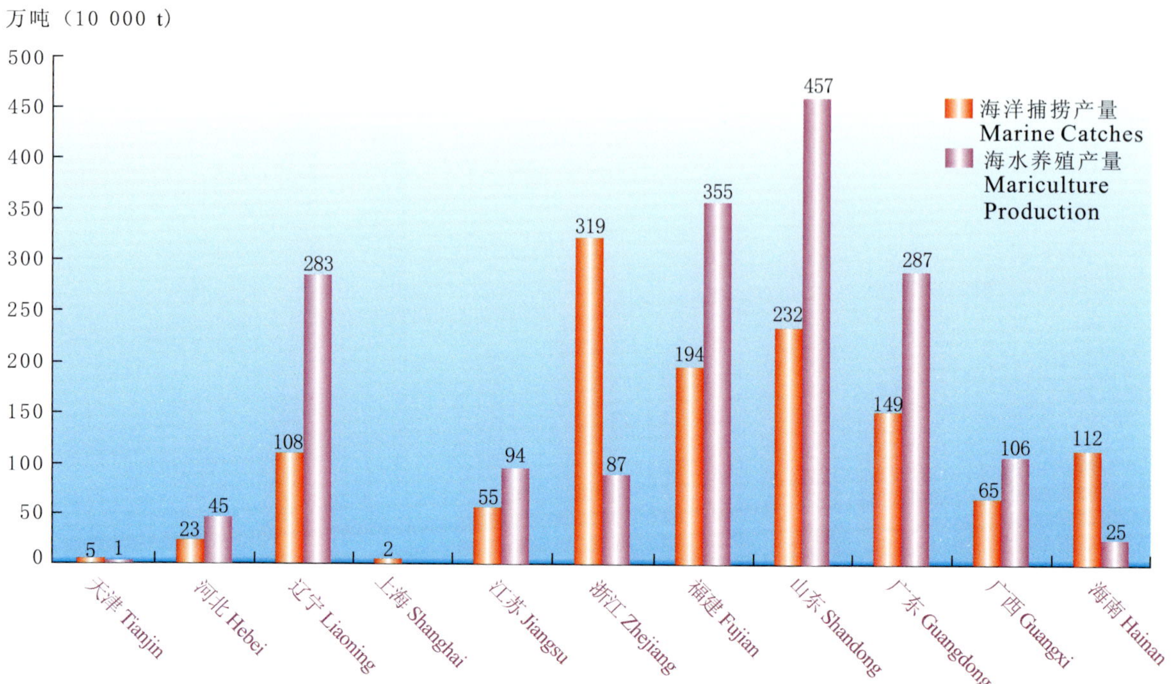

图 6 全国海洋石油和天然气产量

National Output of Offshore Oil and Natural Gas

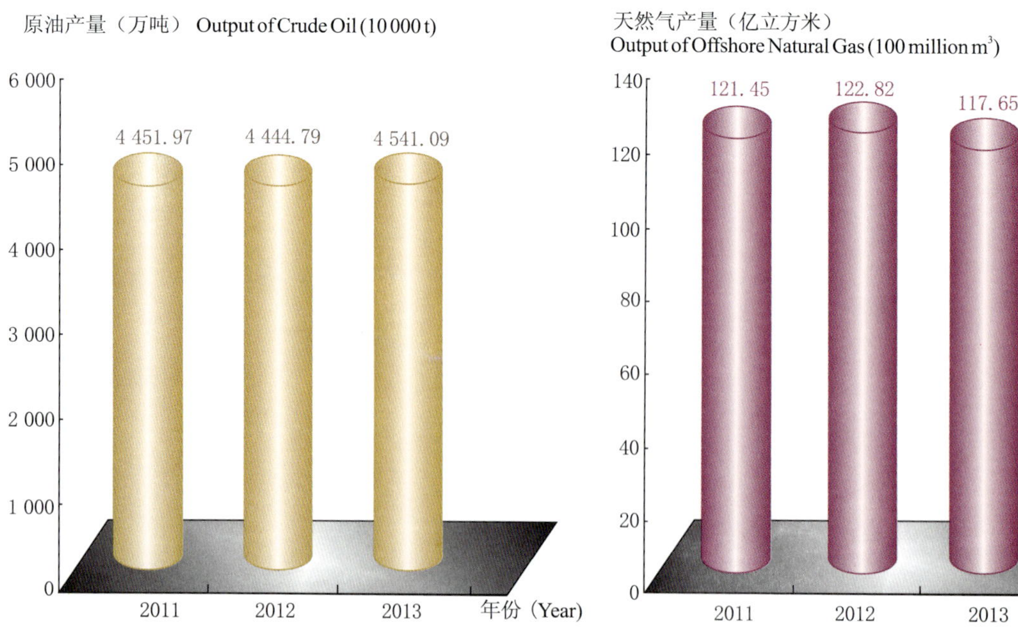

图 7 海洋原油产量占全国原油产量比重
Proportion of Offshore Crude Oil Production in the National Total

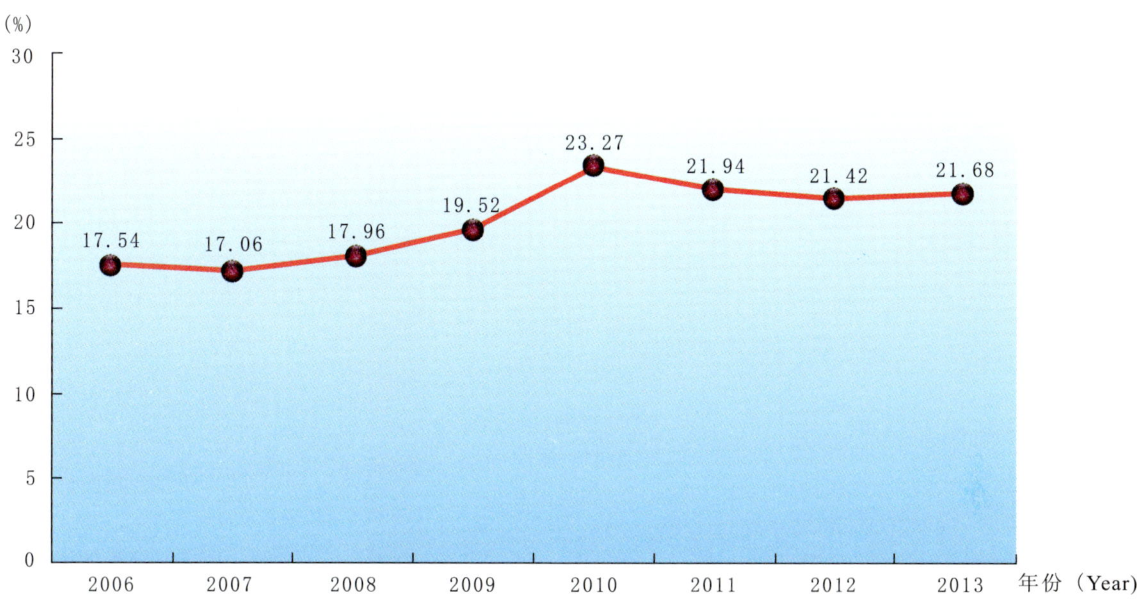

图 8 2013年海洋矿业产量
Production of Marine Mining Industry in 2013

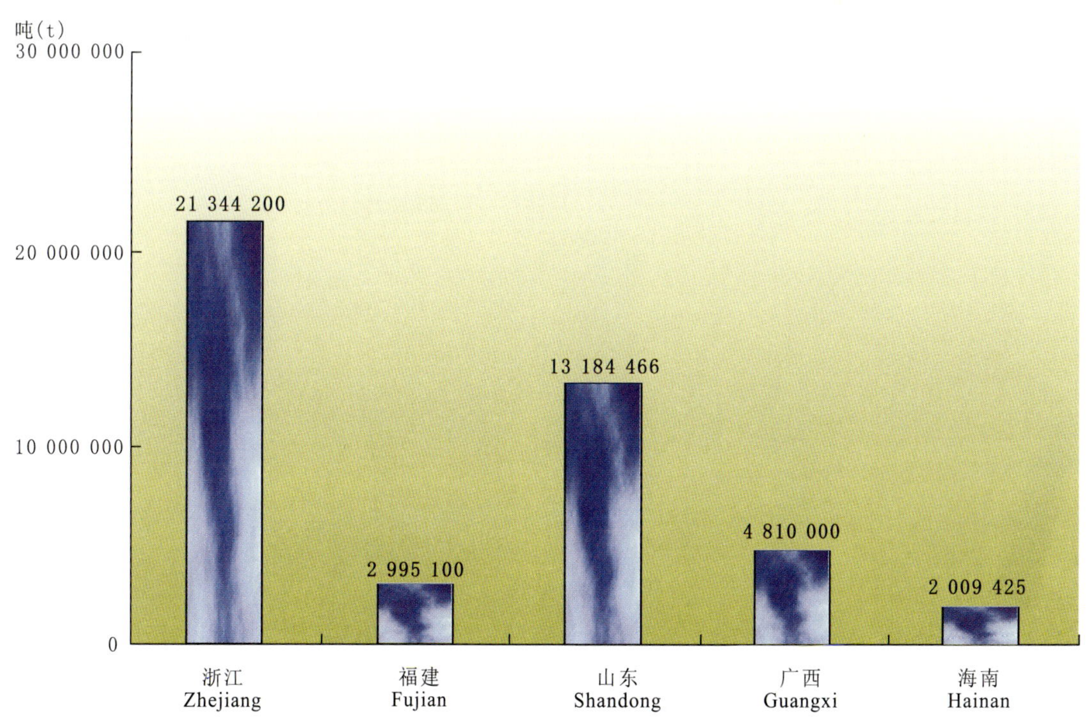

图 9　2013年沿海地区海盐产量

Sea Salt Production by Coastal Regions in 2013

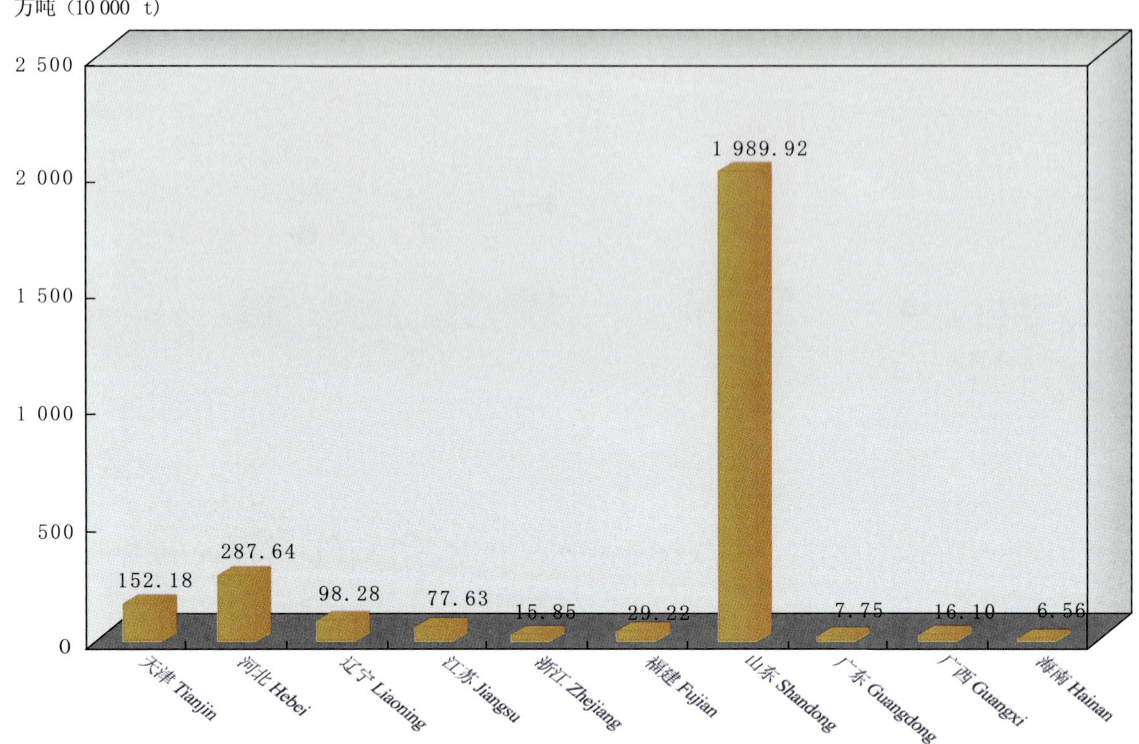

图 10　2013年沿海地区海洋造船完工量

Completed Quantity of Marine Shipbuilding in the Coastal Region in 2013

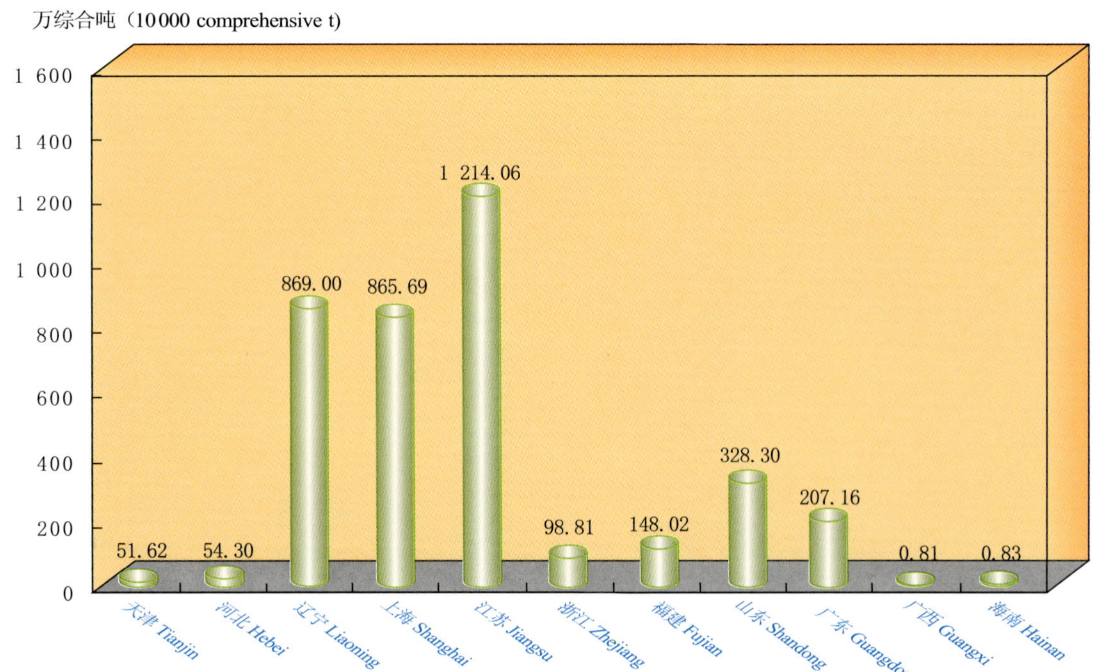

图 11　2013年沿海地区海洋货物周转量
Maritime Goods Turnover Volume by Coastal Regions in 2013

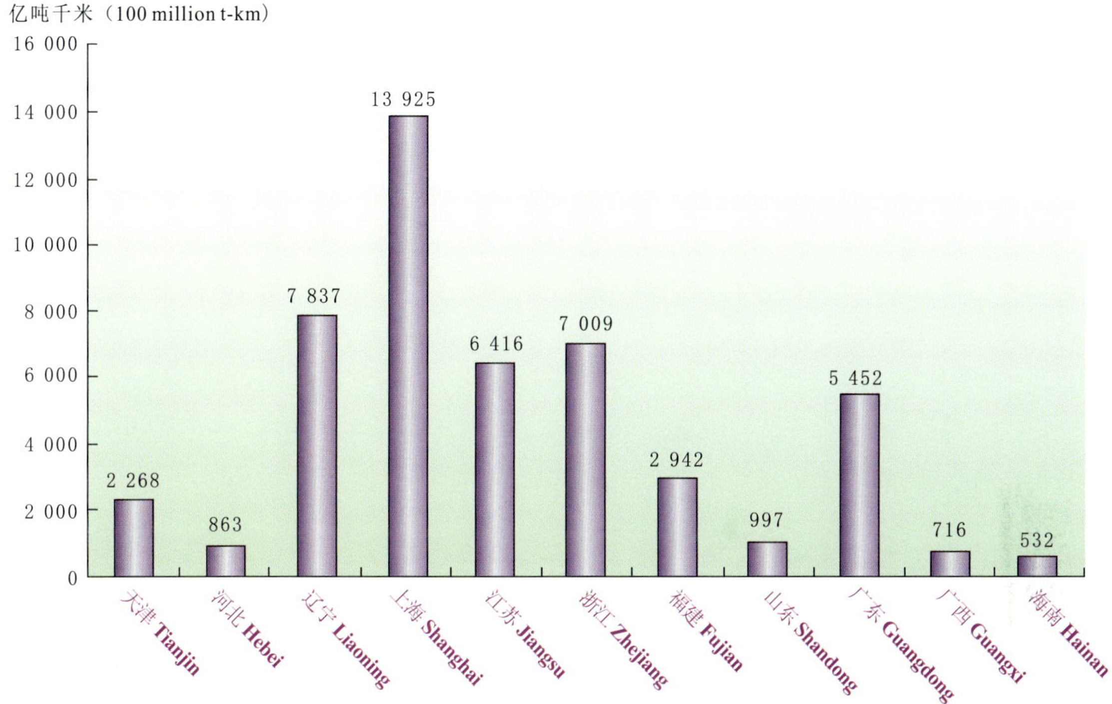

图 12　2013年沿海港口国际标准集装箱吞吐量
International Standardized Containers Handled by Coastal Seaports in 2013

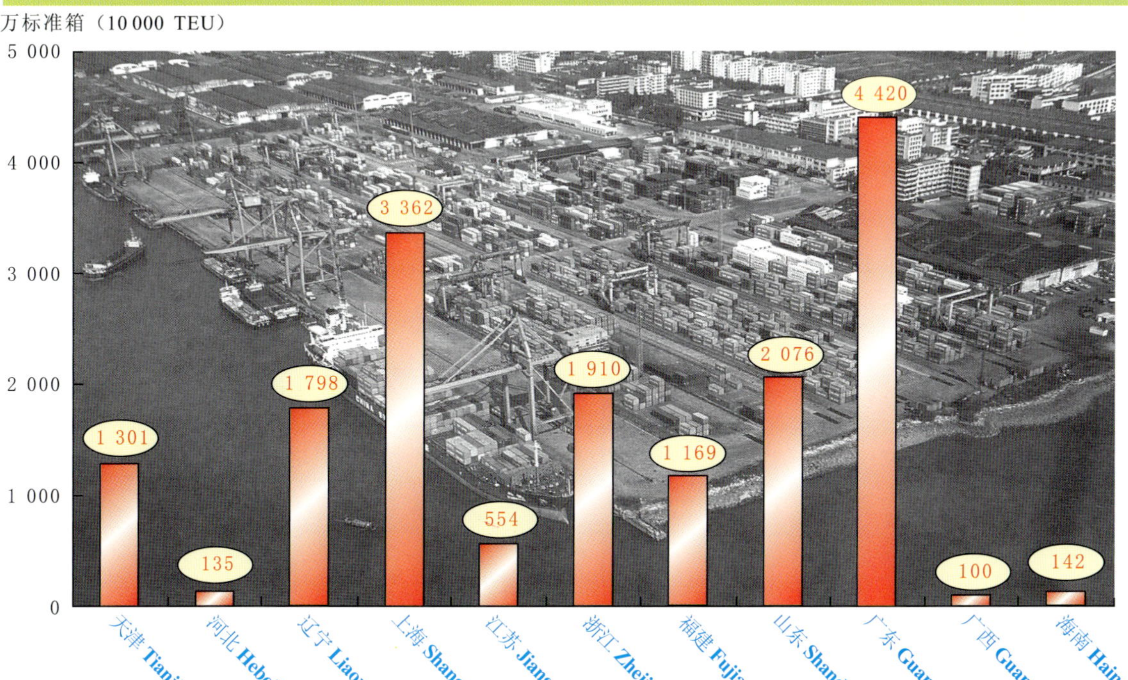

图 13 主要沿海城市国际旅游（外汇）收入

Foreign Exchange Earnings from International Tourism in Major Coastal Cities

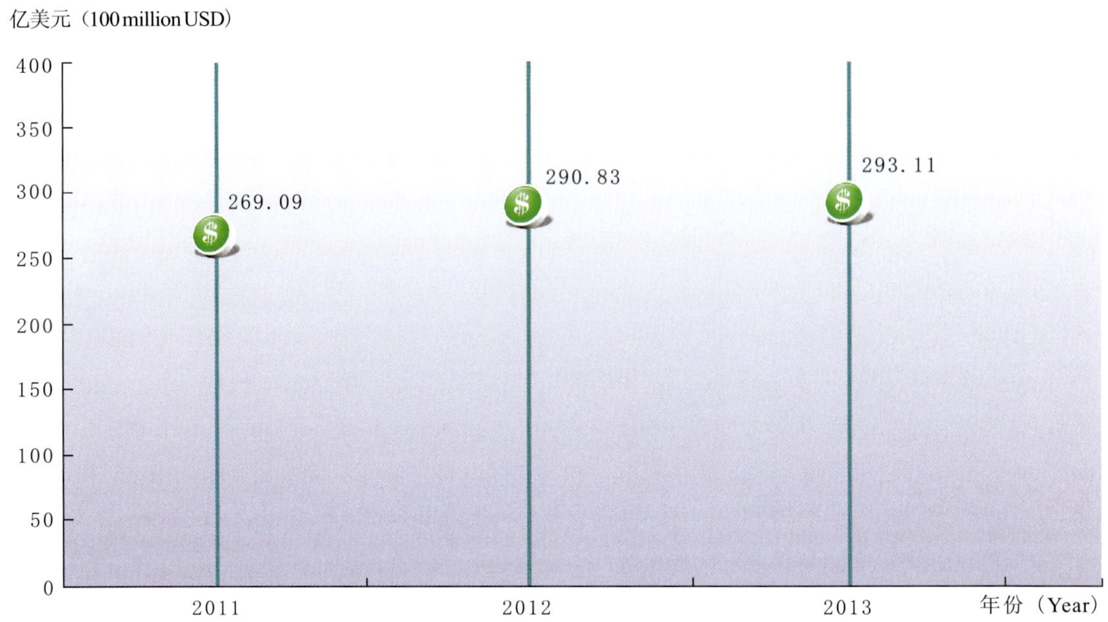

图 14 主要沿海城市接待入境旅游者人数

Number of Oversea Visitor Arrivals in Major Coastal Cities

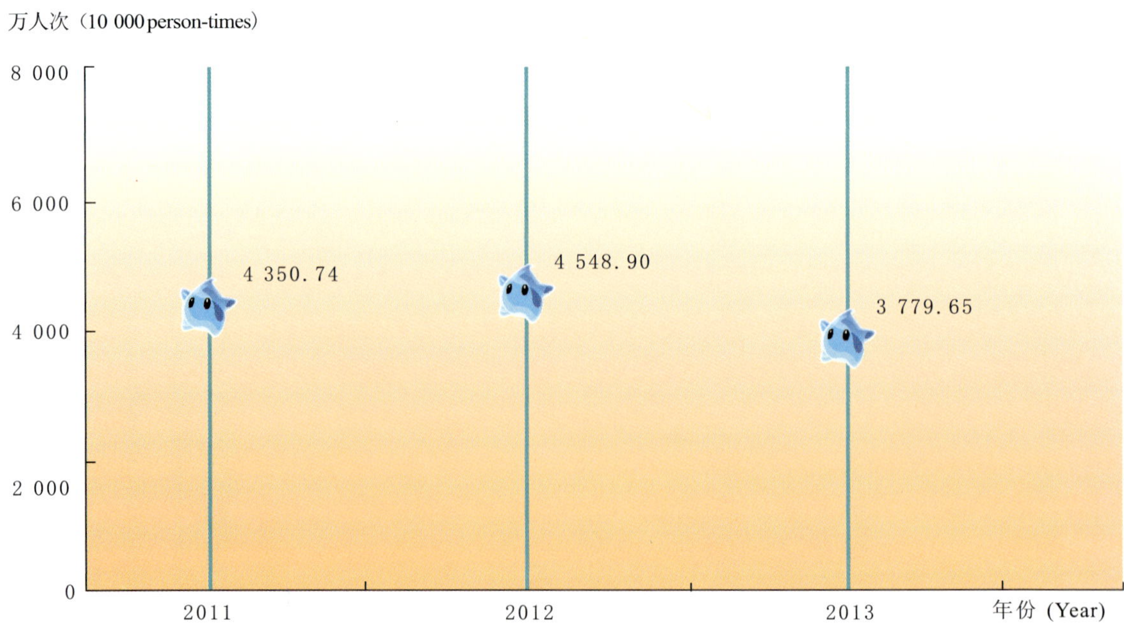

图 15 2013年三大经济区地区生产总值与海洋生产总值
GDP and GOP in the Three Major Economic Zones in 2013

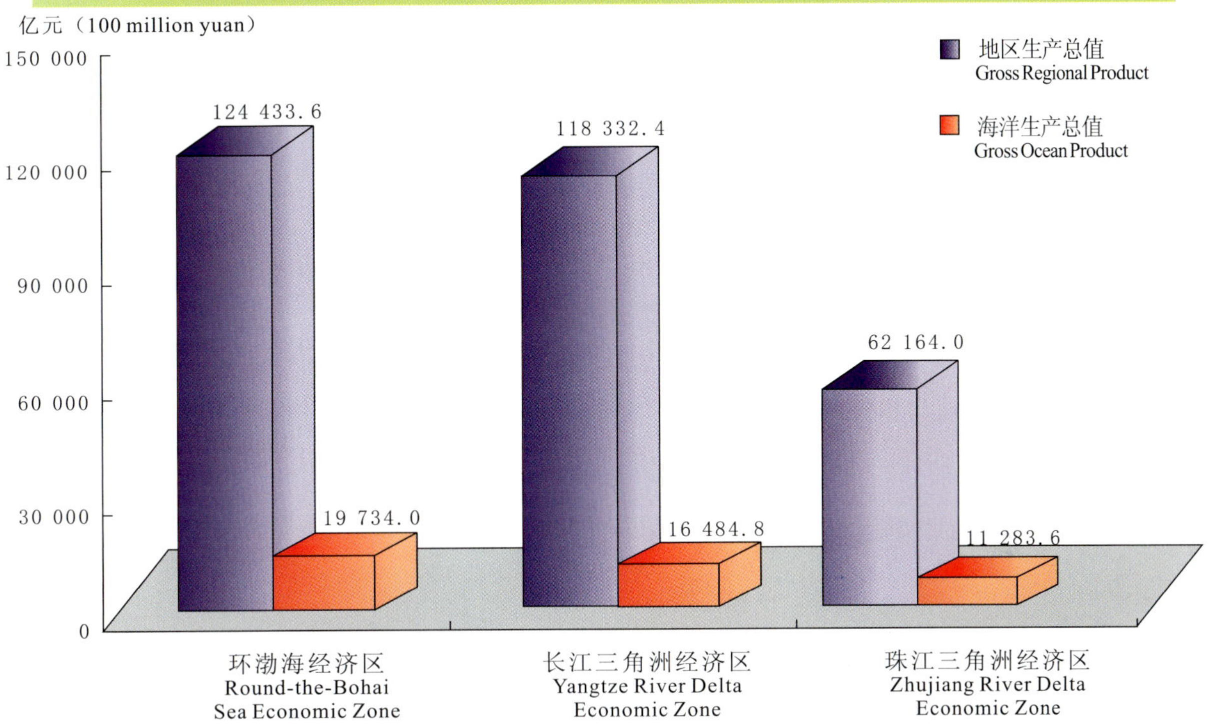

图 16 2013年三大经济区海洋生产总值占全国海洋生产总值比重
Proportion of the GOP of the Three Major Economic Zones in the National GOP in 2013

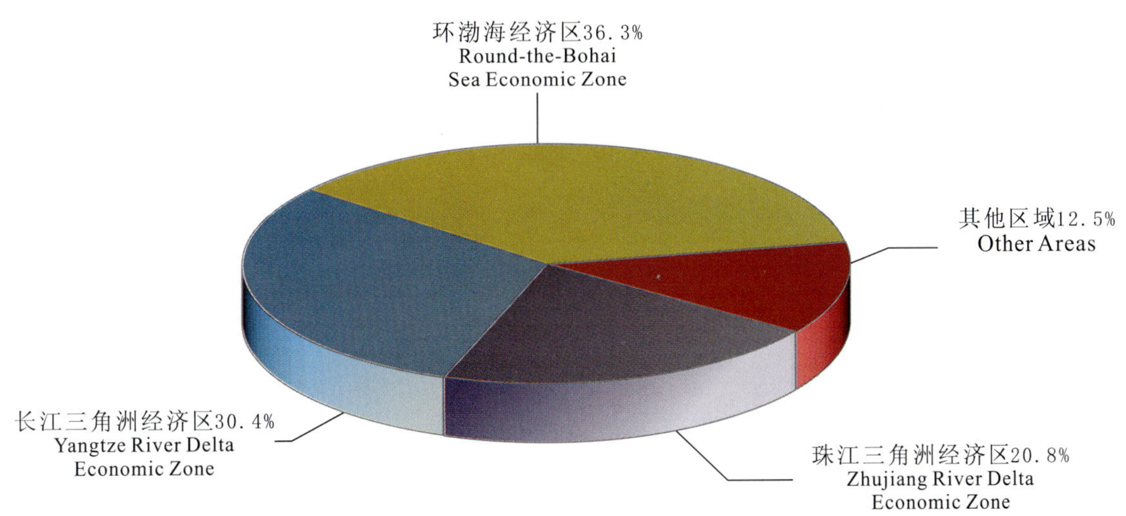

图 17 2013年沿海地区海洋生产总值

Gross Ocean Product by Coastal Regions in 2013

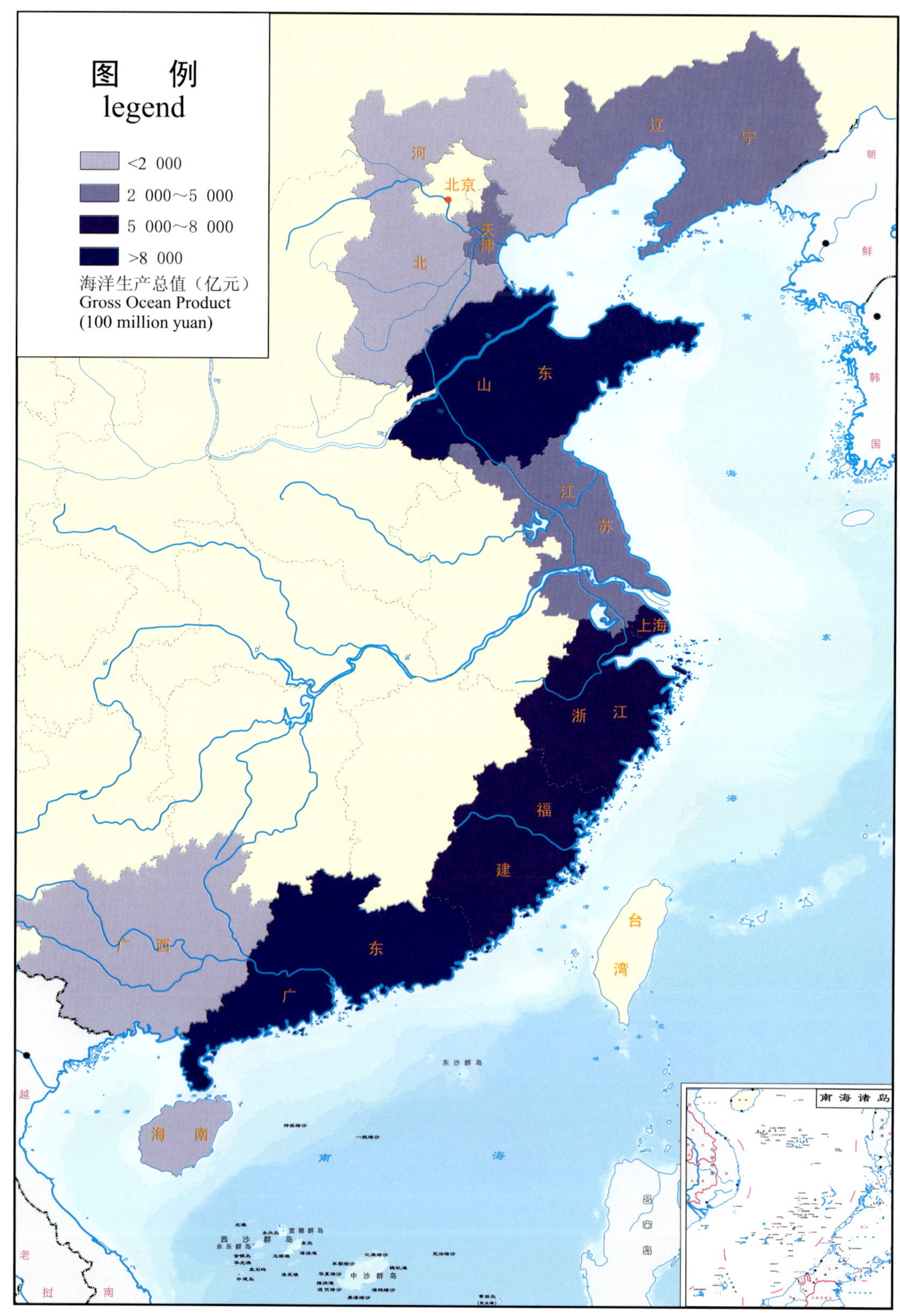

图 18 2013年沿海地区海洋经济贡献
Marine Economic Contributions by Coastal Regions in 2013

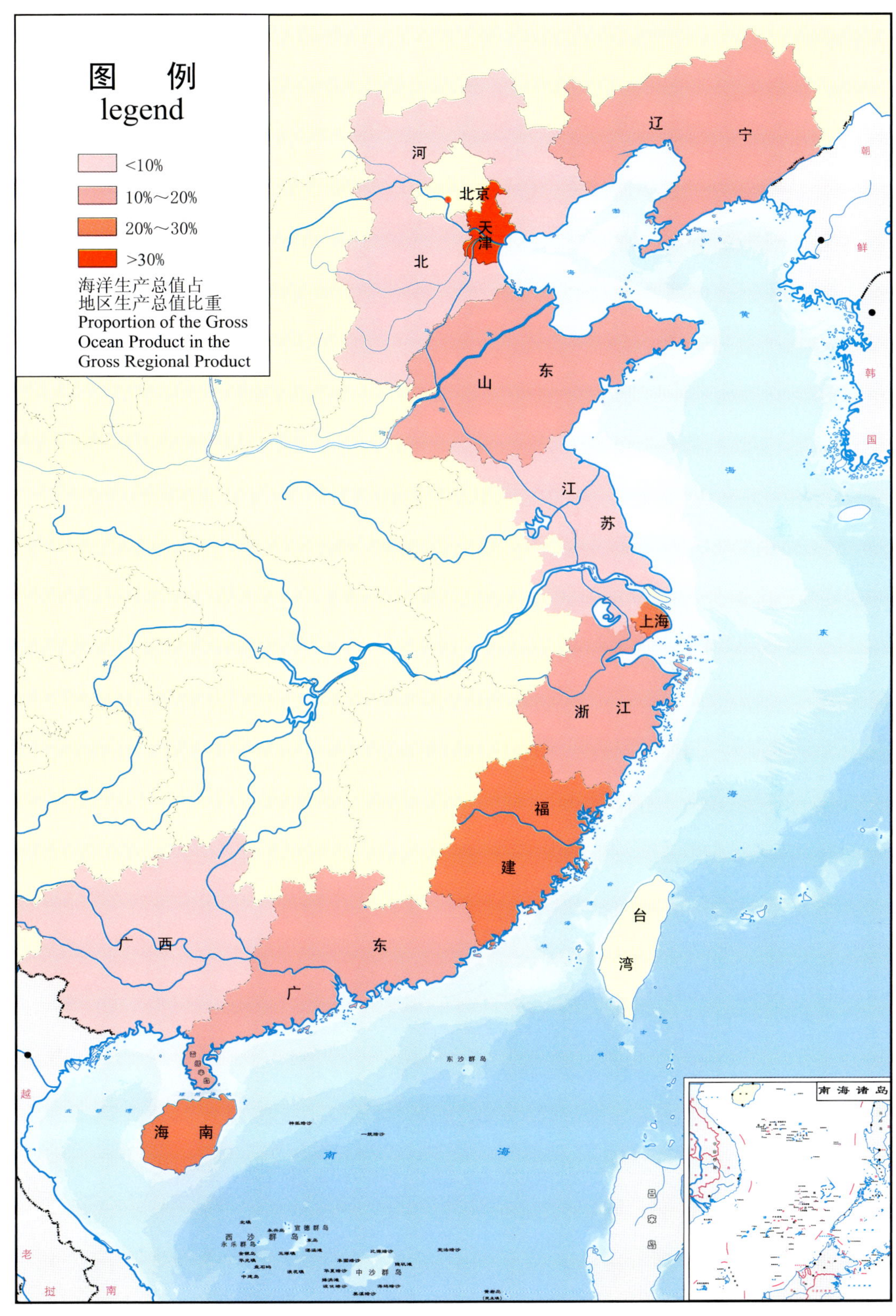

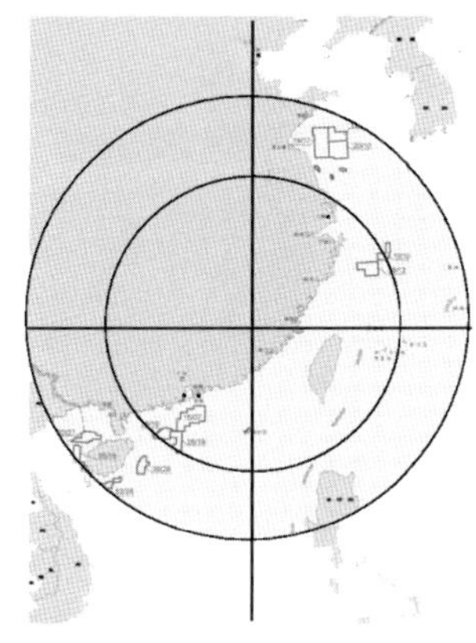

1

综 合 资 料

Integrated Data

1-1　沿海地区行政区划
Administrative Division of Coastal Regions

单位：个　　(number)

沿海地区 Coastal Region	沿海城市 Coastal City	沿海地带 Coastal County (District)			
		合　计 Total	县 County	县级市 County-level City	区 District
合　计 Total	**54**	**236**	**62**	**54**	**120**
天　津 Tianjin	1	1			1
河　北 Hebei	3	11	6	1	4
辽　宁 Liaoning	6	22	4	7	11
上　海 Shanghai	1	5	1		4
江　苏 Jiangsu	3	15	8	4	3
浙　江 Zhejiang	7	35	10	9	16
福　建 Fujian	6	34	11	8	15
山　东 Shandong	7	35	6	13	16
广　东 Guangdong	14	56	10	6	40
广　西 Guangxi	3	8	1	1	6
海　南 Hainan	3	14	5	5	4

注：沿海地带中未包括广东省的东莞、中山和海南省的三沙市所辖地区。

Note: The coastal zone does not includes the regions under the jurisdiction of Dongguan city, Zhongshan city, Guangdong Province and Sansha City, Hainan Province.

1-2 沿海行政区划一览表
Table of Administrative Division of Coastal Regions

沿海地区 Coastal Region	地区代码 Zip Code	沿海城市 Coastal City	地区代码 Zip Code	沿海地带 Coastal County (District)	地区代码 Zip Code
天 津 Tianjin	120000			滨海新区Binhai Xinqu	120116
河 北 Hebei	130000	唐山 Tangshan	130200	丰南区Fengnan Qu	130207
				曹妃甸区Caofeidian Qu	130209
				滦南县Luannan Xian	130224
				乐亭县Leting Xian	130225
		秦皇岛 Qinhuangdao	130300	海港区Haigang Qu	130302
				山海关区Shanhaiguan Qu	130303
				北戴河区Beidaihe Qu	130304
				昌黎县Changli Xian	130322
				抚宁县Funing Xian	130323
		沧州 Cangzhou	130900	海兴县Haixing Xian	130924
				黄骅市Huanghua Shi	130983
辽 宁 Liaoning	210000	大连 Dalian	210200	中山区Zhongshan Qu	210202
				西岗区Xigang Qu	210203
				沙河口区Shahekou Qu	210204
				甘井子区Ganjingzi Qu	210211
				旅顺口区Lüshunkou Qu	210212
				金州区Jinzhou Qu	210213
				长海县Changhai Xian	210224
				瓦房店市Wafangdian Shi	210281
				普兰店市Pulandian Shi	210282
				庄河市Zhuanghe Shi	210283
		丹东 Dandong	210600	振兴区Zhenxing Qu	210603
				东港市Donggang Shi	210681
		锦州 Jinzhou	210700	凌海市Linghai Shi	210781
		营口 Yingkou	210800	鲅鱼圈区Bayuquan Qu	210804
				老边区Laobian Qu	210811
				盖州市Gaizhou Shi	210881
		盘锦 Panjin	211100	大洼县Dawa Xian	211121
				盘山县Panshan Xian	211122

1-2 续表1 continued

沿海地区 Coastal Region	地区代码 Zip Code	沿海城市 Coastal City	地区代码 Zip Code	沿海地带 Coastal County (District)	地区代码 Zip Code
		葫芦岛 Huludao	211400	连山区Lianshan Qu	211402
				龙港区Longgang Qu	211403
				绥中县Suizhong Xian	211421
				兴城市Xingcheng Shi	211481
上海 Shanghai	310000			宝山区Baoshan Qu	310113
				浦东新区Pudong Xinqu	310115
				金山区Jinshan Qu	310116
				奉贤区Fengxian Qu	310120
				崇明县Chongming Xian	310230
江苏 Jiangsu	320000	南通 Nantong	320600	通州区Tongzhou Qu	320612
				海安县Hai'an Xian	320621
				如东县Rudong Xian	320623
				启东市Qidong Shi	320681
				海门市Haimen Shi	320684
		连云港 Lianyungang	320700	连云区Lianyun Qu	320703
				新浦区Xinpu Qu	320705
				赣榆县Ganyu Xian	320721
				灌云县Guanyun Xian	320723
				灌南县Guannan Xian	320724
		盐城 Yancheng	320900	响水县Xiangshui Xian	320921
				滨海县Binhai Xian	320922
				射阳县Sheyang Xian	320924
				东台市Dongtai Shi	320981
				大丰市Dafeng Shi	320982
浙江 Zhejiang	330000	杭州 Hangzhou	330100	滨江区Binjiang Qu	330108
				萧山区Xiaoshan Qu	330109
		宁波 Ningbo	330200	海曙区Haishu Qu	330203
				江东区Jiangdong Qu	330204
				江北区Jiangbei Qu	330205
				北仑区Beilun Qu	330206
				镇海区Zhenhai Qu	330211
				鄞州区Yinzhou Qu	330212
				象山县Xiangshan Xian	330225
				宁海县Ninghai Xian	330226
				余姚市Yuyao Shi	330281
				慈溪市Cixi Shi	330282
				奉化市Fenghua Shi	330283

1-2 续表2 continued

沿海地区 Coastal Region	地区代码 Zip Code	沿海城市 Coastal City	地区代码 Zip Code	沿海地带 Coastal County (District)	地区代码 Zip Code
		温州 Wenzhou	330300	龙湾区Longwan Qu	330303
				瓯海区Ouhai Qu	330304
				洞头县Dongtou Xian	330322
				平阳县Pingyang Xian	330326
				苍南县Cangnan Xian	330327
				瑞安市Rui'an Shi	330381
				乐清市Yueqing Shi	330382
		嘉兴 Jiaxing	330400	海盐县Haiyan Xian	330424
				海宁市Haining Shi	330481
				平湖市Pinghu Shi	330482
		绍兴 Shaoxing	330600	柯桥区Keqiao Qu	330603
				上虞区Shangyu Qu	330604
		舟山 Zhoushan	330900	定海区Dinghai Qu	330902
				普陀区Putuo Qu	330903
				岱山县Daishan Xian	330921
				嵊泗县Shengsi Xian	330922
		台州 Taizhou	331000	椒江区Jiaojiang Qu	331002
				路桥区Luqiao Qu	331004
				玉环县Yuhuan Xian	331021
				三门县Sanmen Xian	331022
				温岭市Wenling Shi	331081
				临海市Linhai Shi	331082
福建 Fujian	350000	福州 Fuzhou	350100	马尾区Mawei Qu	350105
				连江县Lianjiang Xian	350122
				罗源县Luoyuan Xian	350123
				平潭县Pingtan Xian	350128
				福清市Fuqing Shi	350181
				长乐市Changle Shi	350182
		厦门 Xiamen	350200	思明区Siming Qu	350203
				海沧区Haicang Qu	350205

1-2 续表3 continued

沿海地区 Coastal Region	地区代码 Zip Code	沿海城市 Coastal City	地区代码 Zip Code	沿海地带 Coastal County (District)	地区代码 Zip Code
				湖里区Huli Qu	350206
				集美区Jimei Qu	350211
				同安区Tong'an Qu	350212
				翔安区Xiang'an Qu	350213
		莆田 Putian	350300	城厢区Chengxiang Qu	350302
				涵江区Hanjiang Qu	350303
				荔城区Licheng Qu	350304
				秀屿区Xiuyu Qu	350305
				仙游县Xianyou Xian	350322
		泉州 Quanzhou	350500	丰泽区Fengze Qu	350503
				洛江区Luojiang Qu	350504
				泉港区Quangang Qu	350505
				惠安县Hui'an Xian	350521
				金门县Jinmen Xian	350527
				石狮市Shishi Shi	350581
				晋江市Jinjiang Shi	350582
				南安市Nan'an Shi	350583
		漳州 Zhangzhou	350600	云霄县Yunxiao Xian	350622
				漳浦县Zhangpu Xian	350623
				诏安县Zhao'an Xian	350624
				东山县Dongshan Xian	350626
				龙海市Longhai Shi	350681
		宁德 Ningde	350900	蕉城区Jiaocheng Qu	350902
				霞浦县Xiapu Xian	350921
				福安市Fu'an Shi	350981
				福鼎市Fuding Shi	350982
山东 Shandong	370000	青岛 Qingdao	370200	市南区Shinan Qu	370202
				市北区Shibei Qu	370203
				黄岛区Huangdao Qu	370211
				崂山区Laoshan Qu	370212
				李沧区Licang Qu	370213
				城阳区Chengyang Qu	370214
				胶州市Jiaozhou Shi	370281
				即墨市Jimo Shi	370282

1-2 续表4 continued

沿海地区 Coastal Region	地区代码 Zip Code	沿海城市 Coastal City	地区代码 Zip Code	沿海地带 Coastal County (District)	地区代码 Zip Code
		东营 Dongying	370500	东营区Dongying Qu	370502
				河口区Hekou Qu	370503
				垦利县Kenli Xian	370521
				利津县Lijin Xian	370522
				广饶县Guangrao Xian	370523
		烟台 Yantai	370600	芝罘区Zhifu Qu	370602
				福山区Fushan Qu	370611
				牟平区Muping Qu	370612
				莱山区Laishan Qu	370613
				长岛县Changdao Xian	370634
				龙口市Longkou Shi	370681
				莱阳市Laiyang Shi	370682
				莱州市Laizhou Shi	370683
				蓬莱市Penglai Shi	370684
				招远市Zhaoyuan Shi	370685
				海阳市Haiyang Shi	370687
		潍坊 Weifang	370700	寒亭区Hanting Qu	370703
				寿光市Shouguang Shi	370783
				昌邑市Changyi Shi	370786
		威海 Weihai	371000	环翠区Huancui Qu	371002
				文登市Wendeng Shi	371081
				荣成市Rongcheng Shi	371082
				乳山市Rushan Shi	371083
		日照 Rizhao	371100	东港区Donggang Qu	371102
				岚山区Lanshan Qu	371103
		滨州 Binzhou	371600	无棣县Wudi Xian	371623
				沾化县Zhanhua Xian	371624
广东 Guangdong	440000	广州 Guangzhou	440100	荔湾区Liwan Qu	440103
				越秀区Yuexiu Qu	440104
				海珠区Haizhu Qu	440105
				天河区Tianhe Qu	440106
				白云区Baiyun Qu	440111
				黄埔区Huangpu Qu	440112
				番禺区Panyu Qu	440113
				南沙区Nansha Qu	440115
				萝岗区Luogang Qu	440116

1-2 续表5 continued

沿海地区 Coastal Region	地区代码 Zip Code	沿海城市 Coastal City	地区代码 Zip Code	沿海地带 Coastal County (District)	地区代码 Zip Code
		深圳 Shenzhen	440300	罗湖区Luohu Qu	440303
				福田区Futian Qu	440304
				南山区Nanshan Qu	440305
				宝安区Bao'an Qu	440306
				龙岗区Longgang Qu	440307
				盐田区Yantian Qu	440308
		珠海 Zhuhai	440400	香洲区Xiangzhou Qu	440402
				斗门区Doumen Qu	440403
				金湾区Jinwan Qu	440404
		汕头 Shantou	440500	龙湖区Longhu Qu	440507
				金平区Jinping Qu	440511
				濠江区Haojiang Qu	440512
				潮阳区Chaoyang Qu	440513
				潮南区Chaonan Qu	440514
				澄海区Chenghai Qu	440583
				南澳县Nan'ao Xian	440523
		江门 Jiangmen	440700	蓬江区Pengjiang Qu	440703
				江海区Jianghai Qu	440704
				新会区Xinhui Qu	440705
				台山市Taishan Shi	440781
				恩平市Enping Shi	440785
		湛江 Zhanjiang	440800	赤坎区Chikan Qu	440802
				霞山区Xiashan Qu	440803
				坡头区Potou Qu	440804
				麻章区Mazhang Qu	440811
				遂溪县Suixi Xian	440823
				徐闻县Xuwen Xian	440825
				廉江市Lianjiang Shi	440881
				雷州市Leizhou Shi	440882
				吴川市Wuchuan Shi	440883
		茂名 Maoming	440900	茂南区Maonan Qu	440902
				茂港区Maogang Qu	440903
				电白县Dianbai Xian	440923
		惠州 Huizhou	441300	惠城区Huicheng Qu	441302
				惠阳区Huiyang Qu	441303
				惠东县Huidong Xian	441323
		汕尾 Shanwei	441500	城 区Chengqu	441502
				海丰县Haifeng Xian	441521
				陆丰市Lufeng Shi	441581

1-2 续表6 continued

沿海地区 Coastal Region	地区代码 Zip Code	沿海城市 Coastal City	地区代码 Zip Code	沿海地带 Coastal County (District)	地区代码 Zip Code
		阳江 Yangjiang	441700	江城区Jiangcheng Qu	441702
				阳西县Yangxi Xian	441721
				阳东县Yangdong Xian	441723
		东莞 Dongguan	441900		
		中山 Zhongshan	442000		
		潮州 Chaozhou	445100	湘桥区Xiangqiao Qu	445102
				饶平县Raoping Xian	445122
		揭阳 Jieyang	445200	榕城区Rongcheng Qu	445202
				揭东区Jiedong Qu	445221
				惠来县Huilai Xian	445224
广西 Guangxi	450000	北海 Beihai	450500	海城区Haicheng Qu	450502
				银海区Yinhai Qu	450503
				铁山港区Tieshangang Qu	450512
				合浦县Hepu Xian	450521
		防城港 Fangchenggang	450600	港口区Gangkou Qu	450602
				防城区Fangcheng Qu	450603
				东兴市Dongxing Shi	450681
		钦州 Qinzhou	450700	钦南区Qinnan Qu	450702
海南 Hainan	460000	海口 Haikou	460100	秀英区Xiuying Qu	460105
				龙华区Longhua Qu	460106
				美兰区Meilan Qu	460108
		三亚 Sanya	460200	市辖区Shixia Qu	460201
		三沙 Sansha	460300	西沙群岛Xisha Qundao	460321
				南沙群岛Nansha Qundao	460322
				中沙群岛的岛礁及其海域 Reefs of Zhongsha Qundao and Their Sea Areas	460323
		省直辖县 Counties Directly under the Hainan Province Goverment	469000	琼海市Qionghai Shi	469002
				儋州市Danzhou Shi	469003
				文昌市Wenchang Shi	469005
				万宁市Wanning Shi	469006
				东方市Dongfang Shi	469007
				澄迈县Chengmai Xian	469023
				临高县Lingao Xian	469024
				昌江黎族自治县Changjiang Lizu Zizhixian	469026
				乐东黎族自治县 Ledong Lizu Zizhixian	469027
				陵水黎族自治县 Lingshui Lizu Zizhixian	469028

1-3 海洋自然地理
Marine Physical Geography

指　　标	Item	指标值 Data
海洋平均深度　（米）	Average Depth of Sea (m)	961
海洋最大深度　（米）	Maximum Depth of Sea (m)	5 559
岸线总长度　（千米）	Total Length of Coastline (km)	32 000
大陆岸线长度	Length of Continental Coastline	18 000
岛屿岸线长度	Length of Insular Coastline	14 000
＞500m^{2}岛屿数　（个）	Number of Islands ＞500 m^2 each (unit)	7 300
岛屿面积　（万平方千米）	Area of Islands (10 000 km^2)	8
已利用的无居民海岛　（个）	Uninhabited Islands Which Have Been Utilized (unit)	1 900
特殊用途海岛	Islands Used for Special Purposes	1 020
公共服务用岛	Islands Used for Public Service	365
旅游娱乐用岛	Islands Used for Tourism and Recreation	73
农林牧渔业用岛	Islands Used for Agriculture, Forestry, Animal Husbandry and Fishery	340
工业、仓储、交通运输用岛	Islands Used for Industry, Storage, Communications and Transport	49

注：海岛数据来源于《全国海岛保护规划》。

Note: Data on islands are derived from the *National Plan for Island Protection*.

1-4 海区海洋石油天然气储量
Offshore Oil and Gas Reserves in the Sea Area

海 区 Sea Area	海洋石油（万吨） Offshore Oil (10 000 t)		海洋天然气（亿立方米） Natural Gas(100 million m^3)	
	累计探明技术可采储量 Proven Technically Recoverable Reserves in the Aggregate	剩余技术可采储量 Surplus Technically Recoverable Reserves	累计探明技术可采储量 Proven Technically Recoverable Reserves in the Aggregate	剩余技术可采储量 Surplus Technically Recoverable Reserves
合计 Total	**100 040.0**	**49 850.0**	**4 533.3**	**3 312.4**
渤海 Bohai Sea	60 827.0	37 093.0	718.3	495.1
黄海 Huanghai Sea				
东海 East China Sea	1 432.0	994.0	685.3	595.4
南海 South China Sea	37 781.0	11 763.0	3 129.7	2 221.9

注：数据来源于《2013年全国矿产资源储量通报》。

Note: The data come from the *Journal on the National Mineral Resources Reserves in 2013* .

1-5 沿海地区水资源情况
Water Resources by Coastal Regions

地 区 Region	水资源总量 （亿立方米） Total Amount of Water Resources (100 million m^3)	地表 水资源量 Surface Water Resources	地下 水资源量 Groundwater Resources	地表水与地下 水资源重复量 Duplicate Measurement of Surface Water and Groundwater	人均水资源量 （立方米/人） Per Capita Water Resources (m^3/person)
全国总计 National Total	**27 957.9**	**26 839.5**	**8 081.1**	**6 962.7**	**2 059.7**
天 津 Tianjin	14.6	10.8	5.0	1.2	101.5
河 北 Hebei	175.9	76.8	138.8	39.8	240.6
辽 宁 Liaoning	463.2	420.3	139.4	96.5	1 055.2
上 海 Shanghai	28.0	22.8	8.2	3.0	116.9
江 苏 Jiangsu	283.5	202.3	97.2	16.0	357.6
浙 江 Zhejiang	931.3	917.3	207.3	193.3	1 697.2
福 建 Fujian	1 151.9	1 150.7	337.6	336.3	3 062.7
山 东 Shandong	291.7	191.1	172.3	71.7	300.4
广 东 Guangdong	2 263.2	2 253.7	532.5	523.1	2 131.2
广 西 Guangxi	2 057.3	2 056.3	478.1	477.1	4 376.8
海 南 Hainan	502.1	496.5	119.5	113.9	5 636.8

注： 数据来源于《2014中国统计年鉴》。

Note: The data come from the *China Statistical Yearbook 2014*.

1-6 沿海地区湿地面积
Area of Wetlands by Coastal Regions

地 区 Region	湿地总面积 （万公顷） Total Area of Wetlands (10 000 hm^2)	#近（海）及海岸 Inshore Areas and Seashores
合 计 **Total**	**1 246.6**	**579.6**
天 津 Tianjin	29.6	10.4
河 北 Hebei	94.2	23.2
辽 宁 Liaoning	139.5	71.3
上 海 Shanghai	46.5	38.7
江 苏 Jiangsu	282.3	108.8
浙 江 Zhejiang	111.0	69.3
福 建 Fujian	87.1	57.6
山 东 Shandong	173.8	72.9
广 东 Guangdong	175.3	81.5
广 西 Guangxi	75.4	25.9
海 南 Hainan	32.0	20.2

注：数据来源于《2014中国统计年鉴》。

Note: The data come from the *China Statistical Yearbook 2014*.

1-7 红树林各地类面积
Site Classification and Area of Sharpleaf Mangrove (*Rhizophora Apiculata*)

单位：公顷 (hm²)

地区 Region	红树林各地类总面积 Total Site Area of Sharpleaf Mangrove	现有面积 Established	未成林面积 Unestablished	宜林地面积 Suitable for Planting
全国总计 National Total	**82 757.2**	**22 024.9**	**1 884.1**	**58 848.2**
浙 江 Zhejiang	5 452.3	20.6	236.1	5 195.6
福 建 Fujian	13 410.1	615.1	286.4	12 508.6
广 东 Guangdong	32 325.9	9 084.0	981.3	22 260.6
广 西 Guangxi	18 029.2	8 374.9	380.3	9 274.0
海 南 Hainan	13 539.7	3 930.3		9 609.4

注：数据来源于《2013中国统计年鉴》。

Note: The data come from the *China Statistical Yearbook 2013*.

1-8 主要沿海城市气候基本情况
Climate of Major Coastal Cities

城市 City	年平均气温（摄氏度） Annual Average Temperature(℃)	年平均相对湿度（%） Annual Average Relative Humidity (%)	全年降水量（毫米） Annual Precipitation (mm)	全年日照时数（小时） Annual Sunshine Hours (h)
天津 Tianjin	12.8	59	411.5	2 255.2
上海 Shanghai	17.6	68	1 173.4	1 864.7
杭州 Hangzhou	18.0	68	1 520.9	1 665.5
福州 Fuzhou	20.4	72	1 137.5	1 578.4
广州 Guangzhou	21.5	81	2 095.4	1 582.9
海口 Haikou	24.3	82	2 067.0	1 725.7

注：数据来源于《2014中国统计年鉴》。

Note: The data come from the *China Statistical Yearbook 2014*.

主要统计指标解释

1. 沿海地区 即广义的沿海地区，是指有海岸线（大陆岸线和岛屿岸线）的地区，按行政区划分为沿海省、自治区、直辖市。

2. 沿海城市 是指有海岸线的直辖市和地级市（包括其下属的全部区、县和县级市）。

3. 沿海地带 即狭义的沿海地区，是指有海岸线的县、县级市、区（包括直辖市和地级市的区）。

4. 海洋 是海和洋的统称。洋为地球表面上相连接的广大咸水水体的主体部分。海为地球表面相连接的广大咸水水体被陆地、岛礁、半岛包围或分隔的边缘部分。

5. 水资源总量 指评价区内降水形成的地表和地下产水总量，即地表产流量与降水入渗补给地下水量之和，不包括过境水量。

6. 地表水资源量 指评价区内河流、湖泊、冰川等地表水体中可以逐年更新的动态水量，即当地天然河川径流量。

7. 地下水资源量 指评价区内降水和地表水对饱水岩土层的补给量，包括降水入渗补给量和河道、湖库、渠系、渠灌田间等地表水体的入渗补给量。

8. 地表水与地下水资源重复量 指地表水和地下水相互转化的部分，即天然河川径流量中的地下水排泄量和地下水补给量中来源于地表水的入渗补给量。

9. 湿地 指天然或人工、长久或暂时性的沼泽地、泥炭地或水域地带，包括静止或流动、淡水、半咸水、咸水体，低潮时水深不超过6米的水域以及海岸地带地区的珊瑚滩和海草床、滩涂、红树林、河口、河流、淡水沼泽、沼泽森林、湖泊、盐沼及盐湖。

10. 红树林 指生长在热带、亚热带低能海岸潮间带上部，受周期性潮水浸淹，以红树植物为主体的常绿灌木或乔木组成的潮滩湿地木本生物群落。

11. 气温 指空气的温度，我国一般以摄氏度(℃)为单位表示。气象观测的温度表是放在离地面约1.5米处通风良好的百叶箱里测量的，因此，通常说的气温指的是离地面1.5米处百叶箱中的温度。其统计计算方法为：

月平均气温是将全月各日的平均气温相加，除以该月的天数而得；

年平均气温是将12个月的月平均气温累加后除以12而得。

12. 相对湿度 指空气中实际所含水蒸气密度和同温度下饱和水蒸气密度的百分比值。其统计方法与气温相同。

13. 降水量 指从天空降落到地面的液态或固态(经融化后)水，未经蒸发、渗透、流失而在地面上积聚的深度。其统计计算方法为：

月降水量是将全月各日的降水量累加而得；

年降水量是将12个月的月降水量累加而得。

14. 日照时数 指太阳实际照射地面的时间。其统计方法与降水量相同。

Explanatory Notes on Main Statistical Indicators

1. Coastal Region, i.e., the coastal region in a broad sense, refers to the regions with coastlines (continental and island coastlines), which are divided into the coastal provinces, autonomous regions and municipalities directly under the Central Government according to the administrative zoning.

2. Coastal City refers to the municipalities directly under the Central Government and the prefecture-level cities (including all the districts, counties and county-level cities under them).

3. Coastal Zone, i.e., the coastal region in a narrow sense, refers to the counties, county-level cities and districts with coastlines (including the districts under the municipalities directly under the Central Government and the prefecture-level districts).

4. Ocean is the general name for sea and ocean. Ocean refers to the main body of large salt water connected with the earth surface. Sea refers to the edge areas of the salt water on the earth surface that are compartmentalized or surrounded by land, island, reef or peninsula.

5. Total Water Resources refers to total volume of water resources measured as run-off for surface water from rainfall and recharge for groundwater in a given area, excluding transit water.

6. Surface Water Resources refers to total renewable resources which exist in rivers, lakes, glaciers and other collectors from rainfall and are measured as run-off of rivers.

7. Groundwater Resources refers to replenishment of aquifers with rainfall and surface water.

8. Duplicated Measurement between Surface Water and Groundwater refers to the exchange between surface water and groundwater, i.e. run-off of rivers includes some depletion into groundwater while groundwater includes some replenishment from surface water.

9. Wetlands refer to marshland and peat bog, whether natural or man-made, permanent or temporary; water covered areas, whether stagnant or flowing, with fresh or brackish-fresh or salty water that is less than 6 meters deep at low tide; as well as coral beach, weed beach, mud beach, mangrove, river outlet, rivers, fresh-water marshland, marshland forests, lakes, salty bog and salt lakes along the coastal areas.

10. Mangrove refers to evergreen woody plants or plant communities in tropical or sub-tropical zones which live between the sea and the land in areas which are inundated by tides.

11. Temperature refers to the air temperature. China uses centigrade as the unit. The thermometry used for weather observation is put in a breezy shutter, which is 1.5 meters high from the ground. Therefore, the commonly used temperature refers to the temperature in the breezy shutter 1.5 meters away from the ground. The calculation method is as follows:

Monthly average temperature is the summation of average daily temperature of one month divided by the actual days of that particular month.

Annual average temperature is the summation of monthly averages of a year divided by 12 months.

12. Relative Humidity refers to the ratio of actual water vapour pressure to the saturated water vapour density under the current temperature. The calculation method is the same as that of temperature.

13. Volume of Precipitation refers to the deepness of liquid state or solid state (thawed) water falling from the sky to the ground that has not evaporated, infiltrated or run off. The calculation method is as follows:

Monthly precipitation is the summation of daily precipitation of a month.

Annual precipitation is the summation of 12 months precipitation of a year.

14. Sunshine Hours refer to the actual hours of sun irradiating the earth. The calculation method is the same as that of the precipitation.

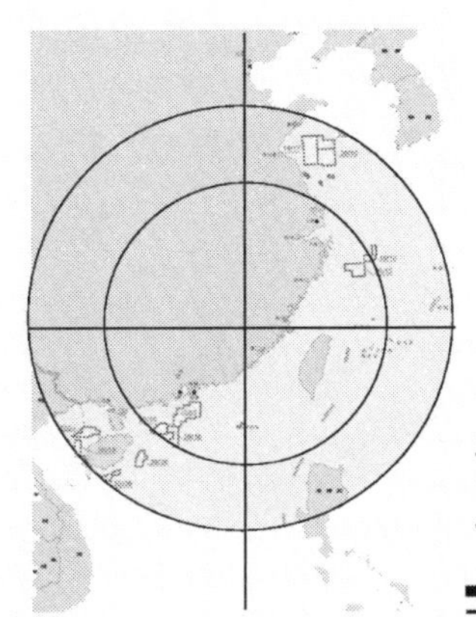

2 海洋经济核算 Marine Economic Accounting

2-1 全国海洋生产总值
National Gross Ocean Product

年 份 Year	海洋生产总值（亿元） Gross Ocean Product (100 million yuan)	第一产业 Primary Industry	第二产业 Secondary Industry	第三产业 Tertiary Industry	海洋生产总值占国内生产总值比重（%） Proportion of the Gross Ocean Product in GDP (%)	海洋生产总值增长速度（%） Growth Rate of the Gross Ocean Product (%)
2001	9 518.4	646.3	4 152.1	4 720.1	8.68	
2002	11 270.5	730.0	4 866.2	5 674.3	9.37	19.8
2003	11 952.3	766.2	5 367.6	5 818.5	8.80	4.2
2004	14 662.0	851.0	6 662.8	7 148.2	9.17	16.9
2005	17 655.6	1 008.9	8 046.9	8 599.8	9.55	16.3
2006	21 592.4	1 228.8	10 217.8	10 145.7	9.98	18.0
2007	25 618.7	1 395.4	12 011.0	12 212.3	9.64	14.8
2008	29 718.0	1 694.3	13 735.3	14 288.4	9.46	9.9
2009	32 277.6	1 857.7	14 980.3	15 439.5	9.47	9.2
2010	39 572.7	2 008.0	18 935.0	18 629.8	9.86	14.7
2011	45 496.0	2 381.9	21 685.6	21 428.5	9.62	9.9
2012	50 045.2	2 670.6	23 469.8	23 904.8	9.64	8.1
2013	54 313.2	2 918.0	24 909.0	26 486.2	9.55	7.6

注：2013年为初步核算数据（以下相关表同）。

Note: The data for 2013 are the preliminary accounts data (The same for the following relevant tables).

2-2 全国海洋生产总值构成
Composition of National Gross Ocean Product

单位：% (%)

年 份 Year	第一产业 Primary Industry	第二产业 Secondary Industry	第三产业 Tertiary Industry
2001	6.8	43.6	49.6
2002	6.5	43.2	50.3
2003	6.4	44.9	48.7
2004	5.8	45.4	48.8
2005	5.7	45.6	48.7
2006	5.7	47.3	47.0
2007	5.4	46.9	47.7
2008	5.7	46.2	48.1
2009	5.8	46.4	47.8
2010	5.1	47.8	47.1
2011	5.2	47.7	47.1
2012	5.3	46.9	47.8
2013	5.4	45.9	48.8

2-3 海洋及相关产业增加值
Added Values of Marine and Related Industries

单位：亿元 (100 million yuan)

年 份 Year	合 计 Total	海洋产业 Marine Industry	主要海洋产业 Major Marine Industry	海洋科研教育管理服务业 Industries of Marine Scientific Research, Education, Management and Service	海洋相关产业 Ocean-related Industries
2001	9 518.4	5 733.6	3 856.6	1 877.0	3 784.8
2002	11 270.5	6 787.3	4 696.8	2 090.5	4 483.2
2003	11 952.3	7 137.7	4 754.4	2 383.3	4 814.6
2004	14 662.0	8 710.1	5 827.7	2 882.5	5 951.9
2005	17 655.6	10 539.0	7 188.0	3 350.9	7 116.6
2006	21 592.4	12 696.7	8 790.4	3 906.4	8 895.6
2007	25 618.7	15 070.6	10 478.3	4 592.3	10 548.0
2008	29 718.0	17 591.2	12 176.0	5 415.2	12 126.8
2009	32 277.6	18 822.0	12 843.6	5 978.4	13 455.6
2010	39 572.7	22 831.0	16 187.8	6 643.1	16 741.7
2011	45 496.0	26 422.0	18 865.2	7 556.8	19 074.1
2012	50 045.2	29 264.4	20 829.9	8 434.6	20 780.8
2013	54 313.2	31 969.5	22 681.1	9 288.4	22 343.7

2-4 海洋及相关产业增加值构成
Composition of the Added Values of Marine and Related Industries

单位：% (%)

年 份 Year	合 计 Total	海洋产业 Marine Industry	主要海洋产业 Major Marine Industry	海洋科研教育管理服务业 Industries of Marine Scientific Research, Education, Management and Service	海洋相关产业 Ocean-related Industries
2001	100.0	60.2	40.5	19.7	39.8
2002	100.0	60.2	41.7	18.5	39.8
2003	100.0	59.7	39.8	19.9	40.3
2004	100.0	59.4	39.7	19.7	40.6
2005	100.0	59.7	40.7	19.0	40.3
2006	100.0	58.8	40.7	18.1	41.2
2007	100.0	58.8	40.9	17.9	41.2
2008	100.0	59.2	41.0	18.2	40.8
2009	100.0	58.3	39.8	18.5	41.7
2010	100.0	57.7	40.9	16.8	42.3
2011	100.0	58.1	41.5	16.6	41.9
2012	100.0	58.5	41.6	16.9	41.5
2013	100.0	58.9	41.8	17.1	41.1

2-5 全国主要海洋产业增加值
Added Values of National Major Marine Industries

主要海洋产业 Major Marine Industry	增加值 （亿元） Added Value (100 million yuan)	比上年增长（%） (按可比价计算) Percentage of Increase Over Last Year (%) (at comparable price)
合计 Total	**22 681.1**	**6.7**
海洋渔业 Marine Fishery Industry	3 872.3	5.5
海洋油气业 Offshore Oil and Natural Gas Industry	1 648.3	0.1
海洋矿业 Marine Mining Industry	49.1	13.7
海洋盐业 Sea Salt Industry	55.5	- 8.1
海洋船舶工业 Marine Shipbuilding Industry	1 182.8	- 7.7
海洋化工业 Marine Chemical Industry	907.6	11.4
海洋生物医药业 Marine Biomedicine Industry	224.3	20.7
海洋工程建筑业 Marine Engineering Architecture Industry	1 680.0	9.4
海洋电力业 Marine Electric Power Industry	86.7	11.9
海水利用业 Marine Seawater Utilization Industry	12.4	9.9
海洋交通运输业 Maritime Communications and Transportation Industry	5 110.8	4.6
滨海旅游业 Coastal Tourism	7 851.4	11.7

2-6 海洋渔业增加值
Added Value of Marine Fishery Industry

单位：亿元 (100 million yuan)

年 份 Year	增加值 Added Value
2001	966.0
2002	1 091.2
2003	1 145.0
2004	1 271.2
2005	1 507.6
2006	1 672.0
2007	1 906.0
2008	2 228.6
2009	2 440.8
2010	2 851.6
2011	3 202.9
2012	3 560.5
2013	3 872.3

2-7 海洋油气业增加值
Added Value of Offshore Oil and Gas Industry

单位：亿元 (100 million yuan)

年 份 Year	增加值 Added Value
2001	176.8
2002	181.8
2003	257.0
2004	345.1
2005	528.2
2006	668.9
2007	666.9
2008	1 020.5
2009	614.1
2010	1 302.2
2011	1 719.7
2012	1 718.7
2013	1 648.3

2-8　海洋矿业增加值
Added Value of Marine Mining Industry

单位：亿元　　(100 million yuan)

年　份 Year	增加值 Added Value
2001	1.0
2002	1.9
2003	3.1
2004	7.9
2005	8.3
2006	13.4
2007	16.3
2008	35.2
2009	41.6
2010	45.2
2011	53.3
2012	45.1
2013	49.1

注：自2008年起部分地区统计矿种增加。

Note: The data from 2008 include added kinds of minerals in some regions.

2-9　海洋盐业增加值
Added Value of Marine Salt Industry

单位：亿元　　(100 million yuan)

年　份 Year	增加值 Added Value
2001	32.6
2002	34.2
2003	28.4
2004	39.0
2005	39.1
2006	37.1
2007	39.9
2008	43.6
2009	43.6
2010	65.5
2011	76.8
2012	60.1
2013	55.5

2-10 海洋船舶工业增加值
Added Value of Marine Shipbuilding Industry

单位：亿元 (100 million yuan)

年 份 Year	增加值 Added Value
2001	109.3
2002	117.4
2003	152.8
2004	204.1
2005	275.5
2006	339.5
2007	524.9
2008	742.6
2009	986.5
2010	1 215.6
2011	1 352.0
2012	1 291.3
2013	1 182.8

2-11 海洋化工业增加值
Added Value of Marine Chemical Industry

单位：亿元 (100 million yuan)

年 份 Year	增加值 Added Value
2001	64.7
2002	77.1
2003	96.3
2004	151.5
2005	153.3
2006	440.4
2007	506.6
2008	416.8
2009	465.3
2010	613.8
2011	695.9
2012	843.0
2013	907.6

注：自2006年起部分地区统计产品品种增加。

Note: The data from 2006 include added kinds of statistical products in some regions.

2-12 海洋生物医药业增加值
Added Value of Marine Biomedicine Industry

单位：亿元 (100 million yuan)

年 份 Year	增加值 Added Value
2001	5.7
2002	13.2
2003	16.5
2004	19.0
2005	28.6
2006	34.8
2007	45.4
2008	56.6
2009	52.1
2010	83.8
2011	150.8
2012	184.7
2013	224.3

2-13 海洋工程建筑业增加值
Added Value of Marine Engineering Architecture

单位：亿元 (100 million yuan)

年 份 Year	增加值 Added Value
2001	109.2
2002	145.4
2003	192.6
2004	231.8
2005	257.2
2006	423.7
2007	499.7
2008	347.8
2009	672.3
2010	874.2
2011	1 086.8
2012	1 353.8
2013	1 680.0

2-14 海洋电力业增加值
Added Value of Marine Electric Power Industry

单位：亿元 (100 million yuan)

年 份 Year	增加值 Added Value
2001	1.8
2002	2.2
2003	2.8
2004	3.1
2005	3.5
2006	4.4
2007	5.1
2008	11.3
2009	20.8
2010	38.1
2011	59.2
2012	77.3
2013	86.7

2-15 海水利用业增加值
Added Value of Seawater Utilization Industry

单位：亿元 (100 million yuan)

年 份 Year	增加值 Added Value
2001	1.1
2002	1.3
2003	1.7
2004	2.4
2005	3.0
2006	5.2
2007	6.2
2008	7.4
2009	7.8
2010	8.9
2011	10.4
2012	11.1
2013	12.4

2-16　海洋交通运输业增加值
Added Value of Marine Communications and Transportation Industry

单位：亿元　　(100 million yuan)

年　份 Year	增加值 Added Value
2001	1 316.4
2002	1 507.4
2003	1 752.5
2004	2 030.7
2005	2 373.3
2006	2 531.4
2007	3 035.6
2008	3 499.3
2009	3 146.6
2010	3 785.8
2011	4 217.5
2012	4 752.6
2013	5 110.8

2-17　滨海旅游业增加值
Added Value of Coastal Tourism

单位：亿元　　(100 million yuan)

年　份 Year	增加值 Added Value
2001	1 072.0
2002	1 523.7
2003	1 105.8
2004	1 522.0
2005	2 010.6
2006	2 619.6
2007	3 225.8
2008	3 766.4
2009	4 352.3
2010	5 303.1
2011	6 239.9
2012	6 931.8
2013	7 851.4

2-18 沿海地区海洋生产总值
Gross Ocean Product by Coastal Regions

地 区 Region	海洋生产总值（亿元） Gross Ocean Product (100 million yuan)	第一产业 Primary Industry	第二产业 Secondary Industry	第三产业 Tertiary Industry	海洋生产总值占沿海地区生产总值比重（%） Proportion of the Gross Ocean Product in the Gross Regional Product（%）
合 计 **Total**	**54 313.2**	**2 918.0**	**24 909.0**	**26 486.2**	**15.8**
天 津 Tianjin	4 554.1	8.7	3 065.7	1 479.7	31.7
河 北 Hebei	1 741.8	77.9	911.4	752.5	6.2
辽 宁 Liaoning	3 741.9	499.6	1 402.7	1 839.6	13.8
上 海 Shanghai	6 305.7	3.9	2 318.0	3 983.8	29.2
江 苏 Jiangsu	4 921.2	225.6	2 432.2	2 263.5	8.3
浙 江 Zhejiang	5 257.9	378.1	2 258.2	2 621.5	14.0
福 建 Fujian	5 028.0	450.6	2 026.2	2 551.2	23.1
山 东 Shandong	9 696.2	715.7	4 593.9	4 386.6	17.7
广 东 Guangdong	11 283.6	192.7	5 352.6	5 738.3	18.2
广 西 Guangxi	899.4	154.0	376.9	368.6	6.3
海 南 Hainan	883.5	211.2	171.3	501.0	28.1

2-19 沿海地区海洋生产总值构成
Composition of Gross Ocean Product by Coastal Regions

单位：% (%)

地 区 Region	海洋生产总值 Gross Ocean Product	第一产业 Primary Industry	第二产业 Secondary Industry	第三产业 Tertiary Industry
合 计 Total	**100.0**	**5.4**	**45.9**	**48.8**
天 津 Tianjin	100.0	0.2	67.3	32.5
河 北 Hebei	100.0	4.5	52.3	43.2
辽 宁 Liaoning	100.0	13.4	37.5	49.2
上 海 Shanghai	100.0	0.1	36.8	63.2
江 苏 Jiangsu	100.0	4.6	49.4	46.0
浙 江 Zhejiang	100.0	7.2	42.9	49.9
福 建 Fujian	100.0	9.0	40.3	50.7
山 东 Shandong	100.0	7.4	47.4	45.2
广 东 Guangdong	100.0	1.7	47.4	50.9
广 西 Guangxi	100.0	17.1	41.9	41.0
海 南 Hainan	100.0	23.9	19.4	56.7

2-20 沿海地区海洋及相关产业增加值
Added Values of Marine and Related Industries by Coastal Regions

单位：亿元 (100 million yuan)

地 区 Region	合 计 Total	海洋产业 Marine Industry	主要海洋产业 Major Marine Industry	海洋科研教育管理服务业 Industries of Marine Scientific Research, Education, Management and Service	海洋相关产业 Ocean-related Industries
合 计 Total	**54 313.2**	**31 969.5**	**22 681.1**	**9 288.4**	**22 343.7**
天 津 Tianjin	4 554.1	2 457.4	2 235.2	222.2	2 096.7
河 北 Hebei	1 741.8	938.6	851.6	87.0	803.2
辽 宁 Liaoning	3 741.9	2 372.8	1 857.2	515.6	1 369.1
上 海 Shanghai	6 305.7	3 757.5	2 335.3	1 422.2	2 548.2
江 苏 Jiangsu	4 921.2	2 820.8	2 054.9	765.9	2 100.5
浙 江 Zhejiang	5 257.9	3 025.9	2 078.2	947.7	2 232.0
福 建 Fujian	5 028.0	2 841.6	2 091.4	750.2	2 186.3
山 东 Shandong	9 696.2	5 726.5	4 232.7	1 493.8	3 969.7
广 东 Guangdong	11 283.6	6 852.3	4 040.3	2 811.9	4 431.4
广 西 Guangxi	899.4	546.3	461.9	84.4	353.2
海 南 Hainan	883.5	630.0	442.5	187.5	253.5

2-21 沿海地区海洋及相关产业增加值构成
Composition of the Added Values of Marine and Related Industries by Coastal Regions

单位：% (%)

地 区 Region	合 计 Total	海洋产业 Marine Industry	主要海洋产业 Major Marine Industry	海洋科研教育管理服务业 Industries of Marine Scientific Research, Education, Management and Service	海洋相关产业 Ocean-related Industries
合 计 Total	**100.0**	**58.9**	**41.8**	**17.1**	**41.1**
天 津 Tianjin	100.0	54.0	49.1	4.9	46.0
河 北 Hebei	100.0	53.9	48.9	5.0	46.1
辽 宁 Liaoning	100.0	63.4	49.6	13.8	36.6
上 海 Shanghai	100.0	59.6	37.0	22.6	40.4
江 苏 Jiangsu	100.0	57.3	41.8	15.6	42.7
浙 江 Zhejiang	100.0	57.6	39.5	18.0	42.4
福 建 Fujian	100.0	56.5	41.6	14.9	43.5
山 东 Shandong	100.0	59.1	43.7	15.4	40.9
广 东 Guangdong	100.0	60.7	35.8	24.9	39.3
广 西 Guangxi	100.0	60.7	51.4	9.4	39.3
海 南 Hainan	100.0	71.3	50.1	21.2	28.7

主要统计指标解释

1. 海洋经济 是开发、利用和保护海洋的各类产业活动以及与之相关联活动的总和。

2. 海洋生产总值 是海洋经济生产总值的简称，指按市场价格计算的沿海地区常住单位在一定时期内海洋经济活动的最终成果，是海洋产业和海洋相关产业增加值之和。

3. 海洋产业 是开发、利用和保护海洋所进行的生产和服务活动，包括海洋渔业、海洋油气业、海洋矿业、海洋盐业、海洋化工业、海洋生物医药业、海洋电力业、海水利用业、海洋船舶工业、海洋工程建筑业、海洋交通运输业、滨海旅游业等主要海洋产业以及海洋科研教育管理服务业。

4. 海洋科研教育管理服务业 是开发、利用和保护海洋过程中所进行的科研、教育、管理及服务等活动，包括海洋信息服务业、海洋环境监测预报服务、海洋保险与社会保障业、海洋科学研究、海洋技术服务业、海洋地质勘查业、海洋环境保护业、海洋教育、海洋管理、海洋社会团体与国际组织等。

5. 海洋相关产业 是指以各种投入产出为联系纽带，与主要海洋产业构成技术经济联系的上下游产业，涉及海洋农林业、海洋设备制造业、涉海产品及材料制造业、涉海建筑与安装业、海洋批发与零售业、涉海服务业等。

6. 海洋三次产业 我国的海洋三次产业划分如下：

海洋第一产业：是指海洋渔业中的海洋水产品、海洋渔业服务业，以及海洋相关产业中属于第一产业范畴的部门。

海洋第二产业：是指海洋渔业中海洋水产品加工、海洋油气业、海洋矿业、海洋盐业、海洋化工业、海洋生物医药业、海洋电力业、海水利用业、海洋船舶工业、海洋工程建筑业，以及海洋相关产业中属于第二产业范畴的部门。

海洋第三产业：是指除海洋第一、第二产业以外的其他行业。第三产业包括：海洋交通运输业、滨海旅游业、海洋科研教育管理服务业，以及海洋相关产业中属于第三产业范畴的部门。

7. 海洋渔业 包括海水养殖、海洋捕捞、海洋渔业服务业和海洋水产品加工等活动。

8. 海洋油气业 是指在海洋中勘探、开采、输送、加工原油和天然气的生产活动。

9. 海洋矿业 包括海滨砂矿、海滨土砂石、海滨地热与煤矿及深海矿物等的采选活动。

10. 海洋盐业 是指利用海水生产以氯化钠为主要成分的盐产品的活动，包括采盐和盐加工。

11. 海洋船舶工业 是指以金属或非金属为主要材料，制造海洋船舶、海上固定及浮动装置的活动，以及对海洋船舶的修理及拆卸活动。

12. 海洋化工业 包括海盐化工、海水化工、海藻化工及海洋石油化工的化工产品生产活动。

13. 海洋生物医药业 是指以海洋生物为原料或提取有效成分，进行海洋药品与海洋保健品的生产加工及制造活动。

14. 海洋工程建筑业 是指在海上、海底和海岸所进行的用于海洋生产、交通、娱乐、防护等用途的建筑工程施工及其准备活动；包括海港建筑、滨海电站建筑、海岸堤坝建筑、海洋隧道桥梁建筑、海上油气田陆地终端及处理设施建造、海底线路管道和设备安装，不包括各部门、各地区的房屋建筑及房屋装修工程。

15. 海洋电力业 是指在沿海地区利用海洋能、海洋风能进行的电力生产活动。不包括沿海地区的火力发电和核力发电。

16. 海水利用业 是指对海水的直接利用和海水淡化活动，包括利用海水进行淡水生产和将海水应用于工业冷却用水和城市生活用水、消防用水等活动，不包括海水化学资源综合利用活动。

17. 海洋交通运输业 是指以船舶为主要工具从事海洋运输以及为海洋运输提供服务的活动，包括远洋旅客运输、沿海旅客运输、远洋货物运输、沿海货物运输、水上运输辅助活动、管道运输业、装卸搬运及其他运输服务活动。

18. 滨海旅游业 是指以海岸带、海岛及海洋各种自然景观、人文景观为依托的旅游经营、服务活动，主要包括：海洋观光游览、休闲娱乐、度假住宿、体育运动等活动。

Explanatory Notes on Main Statistical Indicators

1. Marine Economy is the summation of various types of industrial activities for developing, utilizing and protecting the ocean as well as the activities associated with there.

2. Gross Ocean Product is the short form of the gross output value of ocean economy, referring to the final result of marine economic activities of the permanent units in the coastal region within a given period calculated at the market price, and the sum total of the added values of the marine industries as the ocean-related industries.

3. Marine industry refers to the production as service activities for developing, utilizing and protecting the ocean, including major marine industries such as offshore oil and gas industry, marine mining industry, marine salt industry, marine chemical industry, marine biomedicine industry, marine electric power industry, seawater utilization industry, marine shipbuilding industry, marine engineering construction industry, marine communications and transportation industry, coastal tourism etc. as well as marine scientific research, education, management and service.

4. Marine Scientific Research, Education, Management and Service refer to the activities of scientific research, education, management and service carried out in the process of developing, utilizing and protecting the ocean, including marine information service industry, marine environment monitoring and forecasting service, marine insurance and social security industry, marine scientific research, marine technological service industry, ocean geological prospecting industry, marine environmental protection industry, marine education, marine management, marine social organization and international organizations etc.

5. Ocean-Related Industry refers to the lower and upper reaches enterprises that form a technical and economic link with the major marine industries, with various inputs and outputs as ties, involving marine agriculture and forestry, marine equipment manufacturing, ocean-related building and installation industry, marine wholesale and retail industry, ocean-related service industry etc.

6. Marine Three Industries Chinese marine three industries are divided as follows:

Marine primary industry: refers to the marine aquatic products, marine fishery service industry in the marine fishery as well as the sectors belonging to the primary industry category in the ocean-related industries.

Marine secondary industry: refers to the marine aquatic products processing industry in the marine fishery, offshore oil as gas industry, marine mining industry, marine salt industry, marine chemical industry, marine biomedicine industry, marine electric power industry, seawater utilization industry, marine shipbuilding industry, marine engineering construction industry, as well as the sectors belonging to the category of secondary industry in the ocean-related industries.

Marine tertiary Industry: refers to the industries other than the marine primary and secondary industries, including marine communications and transportation industry, coastal tourism, marine scientific research, education, management and service industry as well as the sectors belonging to the category of tertiary industry in the ocean-related industries.

7. Marine Fishery includes mariculture, marine fishing, marine fishery service industry and marine aquatic products processing, etc.

8. Offshore Oil and Gas Industry refers to the production activities of exploring, exploiting, transporting and processing crude oil and natural gas in the ocean.

9. Ocean Mining Industry includes the activities of extracting and dressing beach placers, beach soil and sand, submarine geothermal energy, and coal mining and deep-sea mining, etc.

10. Marine Salt Industry refers to the activity of producing the salt products with the sodium chloride as the main component by utilizing seawater, including salt extracting and processing.

11. Shipbuilding Industry refers to the activity of building ocean vessels, offshore fixed and floating equipment with metals or non-metals as main materials as well as repairing and dismantling ocean vessels.

12. Marine Chemical Industry includes the production activities of chemical products of sea salt, seawater, sea algal and marine petroleum chemical industries.

13. Marine Biomedicine Industry refers to the production, processing and manufacturing activities of marine medicines and marine health care products by using marine organisms as raw materials or extracting useful components therefrom.

14. Marine Engineering Building Industry refers to the architectural projects construction and its preparations in the sea, at the sea bottom and seacoast for such uses as marine production, transportation, recreation, protection, etc., including constructions of seaports, coastal power stations, coastal dykes, marine tunnels and bridges, land terminals of offshore oil and gas fields as well as building of processing facilities, and installation of submarine pipelines and equipment, but not the projects of house building and renovation.

15. Marine Electric Power Industry refers to the activities of generating electric power in the coastal region by making use of ocean energies and ocean wind energy. It does not include the thermal and nuclear power generation in the coastal area.

16. Seawater Utilization Industry refers to the activities of the direct use of sea water and the seawater desalination, including those of carrying out the production of desalination and applying the seawater as water for industrial cooling, urban domestic water, water for fire fighting etc., but not the activity of the multipurpose use of seawater chemical resources.

17. Marine Communications and Transportation Industry refers to the activities of carrying out and serving the sea transportations with vessels as main vehicles, including ocean-going passagers transportation, coastal passagers transportation, ocean-going cargo transportation, coastal cargo transportation, auxiliary activities of water transportation, pipeline transportation, loading, unloading and transport as well as other transportation service activities.

18. Coastal Tourism refers to the tourist business and service activities with the backing of coastal zone, sea islands as well as a variety of natural and human landscapes of the ocean, mainly including marine sightseeing, living a life of leisure and recreation, going on vocation and getting accommodation, sports, etc.

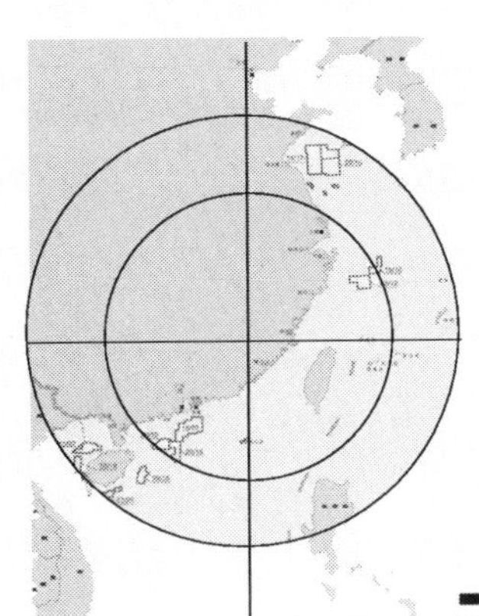

3 主要海洋产业活动

Major Marine Industrial Activities

3-1 全国海洋捕捞养殖产量
National Marine Catches and Mariculture Production

单位：吨 (t)

项　目　Item	2011	2012	2013
远洋捕捞产量 Deep-sea fishing production	**1 147 809**	**1 223 441**	**1 351 978**
海洋捕捞产量 Marine Catches	**12 419 386**	**12 671 891**	**12 643 822**
按品种分 By Species			
鱼类 Fish	8 639 947	8 758 466	8 717 638
甲壳类 Crustacea	2 091 282	2 207 391	2 285 476
贝类 Shellfish	584 078	563 422	547 556
藻类 Algae	27 362	25 728	28 036
头足类 Cephalopoda	695 251	698 909	664 285
其他 Others	381 466	417 975	400 831
海水养殖产量 Mariculture Production	**15 513 292**	**16 438 105**	**17 392 453**
按品种分 By Species			
鱼类 Fish	964 189	1 028 399	1 123 576
甲壳类 Crustacea	1 127 189	1 249 554	1 340 218
贝类 Shellfish	1 153 626	12 084 393	12 728 037
藻类 Algae	1 601 764	1 764 684	1 856 804
其他 Others	276 524	311 075	343 818

3-2 沿海地区海洋捕捞养殖产量

Marine Catches and Mariculture Production by Coastal Regions

单位：吨 (t)

地 区 Region	海洋捕捞产量 Marine Catches	远洋捕捞产量 Deep-Sea Fishing Production	海水养殖产量 Mariculture Production
全国总计 National Total	**12 643 822**	**1 351 978**	**17 392 453**
天 津 Tianjin	53 437	13 027	12 269
河 北 Hebei	230 539		452 270
辽 宁 Liaoning	1 079 259	204 434	2 827 609
上 海 Shanghai	19 639	105 186	
江 苏 Jiangsu	553 787	19 549	938 742
浙 江 Zhejiang	3 192 000	368 186	871 700
福 建 Fujian	1 937 300	230 526	3 548 960
山 东 Shandong	2 315 178	113 062	4 566 350
广 东 Guangdong	1 490 821	63 181	2 870 020
广 西 Guangxi	650 599	2 789	1 056 461
海 南 Hainan	1 121 263		248 072

注：远洋捕捞产量全国总计包括北京7 008吨，中农发集团225 030吨。

Note:Data for the national total deep-sea fishing production includes 7 008 tons from Beijing and 225 030 tons from the China National Agricultural Development Group Co., Ltd..

3-3 沿海地区海洋原油产量
Output of Offshore Crude Oil by Coastal Regions

单位：万吨 (10 000 t)

地 区 Region	2011	2012	2013
合 计 Total	**4 451.97**	**4 444.79**	**4 541.09**
天 津 Tianjin	2 770.20	2 680.34	2 634.68
河 北 Hebei	229.05	237.77	239.87
辽 宁 Liaoning	10.75	14.25	11.29
上 海 Shanghai	16.10	14.68	18.65
山 东 Shandong	257.15	275.00	291.30
广 东 Guangdong	1 168.72	1 222.75	1 345.30

3-4 沿海地区海洋天然气产量
Output of Offshore Natural Gas by Coastal Regions

单位：万立方米 (10 000 m^3)

地 区 Region	2011	2012	2013
合 计 Total	**1 214 519**	**1 228 188**	**1 176 455**
天 津 Tianjin	213 719	246 705	261 682
河 北 Hebei	50 987	55 570	67 201
辽 宁 Liaoning	2 370	1 580	1 410
上 海 Shanghai	71 789	88 044	79 702
山 东 Shandong	12 280	12 521	12 911
广 东 Guangdong	863 374	823 768	753 549

3-5 海洋原油出口量及创汇额
Export Volume of and Foreign-Exchange Earnings from Offshore Crude Oil by Coastal Regions

单位：万吨，万美元 (10 000 t, 10 000 USD)

地 区 Region	2011		2012		2013	
	出口量 Export Volume	创汇额 Foreign-Exchange Earnings	出口量 Export Volume	创汇额 Foreign-Exchange Earnings	出口量 Export Volume	创汇额 Foreign-Exchange Earnings
合 计 **Total**	**39.99**	**29 766**	**30.20**	**23 374**	**16.74**	**11 779**
天 津 Tianjin	25.25	19 009	18.30	13 722	16.74	11 779
广 东 Guangdong	14.74	10 757	11.90	9 652		

3-6 海洋原油产量、出口量占全国原油产量、出口量比重
Proportion of Offshore Crude Oil Production and Export Volume in the National Total

年 份 Year	海洋原油产量 占全国原油产量比重（%） Proportion of Offshore Crude Oil Production in the National Total (%)	海洋原油出口量占 全国原油出口量比重（%） Proportion of Offshore Crude Oil Export Volume in the National Total (%)
2001	13.07	46.26
2002	14.40	54.95
2003	15.01	60.77
2004	16.16	83.37
2005	17.51	84.18
2006	17.54	89.53
2007	17.06	71.53
2008	17.96	76.46
2009	19.52	26.61
2010	23.27	24.38
2011	21.94	15.87
2012	21.42	12.43
2013	21.68	10.33

3-7 沿海地区海洋矿业产量
Output of Marine Mining Industry by Coastal Regions

单位：吨 (t)

地 区 Region	产 量 Output 2011	2012	2013
合 计 Total	**42 310 427**	**43 512 901**	**44 343 191**
浙 江 Zhejiang	27 498 400	27 269 800	21 344 200
福 建 Fujian	2 598 700	2 645 700	2 995 100
山 东 Shandong	9 453 747	10 184 021	13 184 466
广 西 Guangxi	260 000	356 600	4 810 000
海 南 Hainan	2 499 580	3 056 780	2 009 425

3-8 沿海地区海盐产量
Sea Salt Production by Coastal Regions

单位：万吨 (10 000 t)

地　区 Region	海盐产量 Output of Sea Salt		
	2011	2012	2013
合　计 **Total**	**3 322.42**	**2 986.42**	**2 681.13**
天　津 Tianjin	181.00	169.95	152.18
河　北 Hebei	402.25	334.69	287.64
辽　宁 Liaoning	133.62	117.39	98.28
江　苏 Jiangsu	94.25	78.10	77.63
浙　江 Zhejiang	13.99	10.88	15.85
福　建 Fujian	48.50	27.04	29.22
山　东 Shandong	2 418.63	2 219.10	1 989.92
广　东 Guangdong	15.80	8.62	7.75
广　西 Guangxi	3.63	16.43	16.10
海　南 Hainan	10.75	4.22	6.56

3-9 沿海地区海洋化工产品产量
Output of Marine Chemical Products by Coastal Regions

单位：吨 (t)

地 区 Region	产品产量[1] Output		
	2011	2012	2013
合 计 Total	**10 890 465**	**17 450 267**	**19 095 811**
天 津 Tianjin	1 543 708	1 610 000	1 641 323
河 北 Hebei	73 960	1 046 405[2]	7 942[2]
辽 宁 Liaoning	892 629	1 020 284	1 021 941
江 苏 Jiangsu	1 762 042	2 090 567	1 933 766
浙 江 Zhejiang	619 852	1 001 170	1 022 884
福 建 Fujian	752 636	1 462 106	1 652 964
山 东 Shandong	4 580 238	8 414 735	10 919 468
广 东 Guangdong	665 400[2]	805 000[2]	895 523[2]

注：①数据为沿海地区部分海洋化工企业产品汇总数据；②为中国盐业总公司数据。

Note: ① The data are collected from the products of part of the chemical enterprises in the coastal region.
② The data are from the China Salt Industry Corporation.

3-10 沿海地区海洋生物医药产品产量*

Production of the Marine Biomedicine Industry by Coastal Regions*

产品名称 Name	计量单位 Unit	产品产量 Output
海参肽胶囊 Sea Cucumber Peptide Capsule	吨 (t)	2.70
螺旋藻粉 Spirulina Powder	吨 (t)	459.00
螺旋藻胶囊 Spirulina Capsule	万粒 (10 000 pellets)	750.00
海藻酸钠 Sodium Algrnic Acid	万吨 (10 000 tons)	1.78
藻酸双酯钠片 Alginic Acid Diadipose Sodium Pill	万片 (10 000 pills)	271.00
藻酸双酯钠（针剂） Alginic Acid Diadipose Sodium Drops	万支 (10 000 bottles)	33.00
鲨鱼肝油胶丸 Shark Liver Oil Pill	万粒 (10 000 pellets)	6 130.90
金枪鱼油胶丸 Tuna Oil Pill	万粒 (10 000 pellets)	1 439.80
金枪鱼油 Tuna Oil	瓶 (bottle)	3 923.00
南海岸鱼油 Southern Coast Fish Oil	万盒 (10 000 cases)	34.23
海藻植物胶囊 Algae Plant Capsule	吨 (t)	489.60
蚝贝钙片 Oyster Shell Calcium Tablets	万片 (10 000 pills)	21 637.94
鱼油软胶囊 Fish Oil Soft Capsule	万瓶 (10 000 bottles)	9.36
鲑鱼油软胶囊 Salmon Oil Soft Capsule	瓶 (bottle)	6 447.00
鱼油磷脂软胶囊 Fish Oil Phospholipid Soft Capsule	万瓶 (10 000 cases)	3.46
鱼肝油 Cod-Liver Oil	万瓶 (10 000 bottles)	300.00
鱼肝油乳 Cod-Liver Oil Milk	万瓶 (10 000 bottles)	226.89
鱼肝油系列产品 Cod-Liver Oil Series Products	箱 (box)	2 100.00
鱼肝油系列产品 Cod-Liver Oil Series Products	吨 (t)	2 020.00
鲨鱼硫酸软骨素 Shark Sulphate Chondroitin	吨 (t)	471.70

3-10 续表1 continued

产品名称 Name	计量单位 Unit	产品产量 Output
海蛇痹宁胶囊 Sea Snake Anti-rheumatic Capsule	万粒 (10 000 pellets)	474.52
维生素AD胶丸 Vitamin AD Capsule	万粒 (10 000 pellets)	47 975.42
维生素AD滴剂 Vitamin AD Drops	万支 (10 000 bottles)	9 821.50
维生素E胶丸 Vitamin E Capsule	万粒 (10 000 pellets)	32 991.10
琼脂 Agar	吨 (t)	2 200.00
海珠喘息定片 Haizhu Methoxyphenamine Pill	万片 (10 000 pills)	19 588.37
珍珠粉末 Pearl Powder	万支 (10 000 bottles)	359.80
金牡感冒片 Jinmu Cold Cure Pill	万片 (10 000 pills)	2 352.34
多烯酸乙酯胶囊 Ethyl Polyenoic Acid Capsule	万瓶 (10 000 bottles)	286.63
润生软胶囊 Runsheng soft Capsule	万瓶 (10 000 bottles)	10.01
几丁聚糖胶囊 Chitosan Capsule	万盒 (10 000 cases)	12.00
氨糖美辛肠溶片 Glucosamine Indometacin Enteric-coated Tablets	万片 (10 000 pills)	585.53
去甲斑蝥素片 Demethylcantharidin Tablets	万片 (10 000 pills)	52.80
浓缩鱼油 Concentrated Fish Oil	吨 (t)	1 888.00
甘糖酯 Mannose Easter	万片 (10 000 pills)	322.00
珠珀惊风散 Zhubo Convulsions Powder	万瓶 (10 000 bottles)	491.00
岩藻黄素 Fucoxanthin	吨 (t)	9.10
虾青素 Astaxanthin	吨 (t)	0.51
脑元神软胶囊 Naoyuanshen Soft Capsule	万粒 (10 000 pellets)	267.23
海麟舒肝胶囊 Hailin Liver-Soothing Capsule	万粒 (10 000 pellets)	27.00
海威口服液 Haiwei Oral Liquid	吨 (t)	428.00

3-10 续表2 continued

产品名称 Name	计量单位 Unit	产品产量 Output
珍珠胎囊口服液 Pearl Capsule Oral Liquid	万支 (10 000 bottles)	415.00
麝珠明目滴眼液 Shezhu Eye-brightening Drops	万瓶 (10 000 bottles)	189.00
卡拉胶 Carrageenan	吨 (t)	7 126.00
角鲨烯 Houndfish Alkene	万瓶 (10 000 bottles)	1.20
角鲨烯胶囊 Houndfish Alkene Capsule	万粒 (10 000 pellets)	413.73
鱼蛋白蔷薇片 Fish Protein Rose Tablet	万瓶 (10 000 bottles)	1.36
鱼蛋白粉 Fish Albumen Powder	吨 (t)	4.55
胶原蛋白 Collagen Albumen	吨 (t)	279.70
鱼油微囊 Fish Oil Capsule	吨 (t)	3 500.00
鱼胶原蛋白肽 Fish Collagen Peptide	万粒 (10 000 pellets)	63.32
海狗油软胶囊 Fur Seal Oil Soft Capsule	万瓶 (10 000 bottles)	2.62
八宝惊风散 Babao Infantile Convulsions Powder	万盒 (10 000 cases)	673.80
甘露醇 Mannitol	吨 (t)	18.00
碘 Iodine	吨 (t)	3 380.00
润科DHA粉剂、油剂 Runke DHA Powder, oil	吨 (t)	1 600.00
润科ARA粉剂、油剂 Runke ARA Powder,oil	吨 (t)	1 000.00
南海岸鳗钙系列产品 South Coast Eel-calcium	万盒 (10 000 cases)	220.00
伤科接骨片 Traumatologic Osteopathic Tablets	吨 (t)	254.43
肤疹宁软膏 Rash Cream	万支 (10 000 bottles)	5.50
康体通胶囊 Kangtitong Capsule	吨 (t)	28.00
鲨鱼软骨罐头 Shark Cartilage Can	罐 (can)	70 000.00

注：数据为沿海地区部分海洋生物医药产品汇总数据。

Note: The data are collected from the products of part of the marine biomedicine enterprices in the coastal region.

3-11 沿海地区海洋修造船完工量
Production of the Marine Shipbuilding Industry by Coastal Regions

部门和地区 Sector and Region	修船完工量（艘） Ships Repaired (unit)	造船完工量 Ships Built	
		艘 (unit)	万综合吨 Comprehensive Tonnages (10 000 t)
总　计　Total	**13 899**	**2 820**	**3 838.60**
其　中: Including:			
中船工业集团公司 CSSC	439	152	1 064.46
中船重工集团公司 CSIC	544	95	790.54
按地区分: By Regions:			
天　津 Tianjin	152	37	51.62
河　北 Hebei	182	8	54.30
辽　宁 Liaoning	240	167	869.00 ①
上　海 Shanghai	1 269	88	865.69
江　苏 Jiangsu	2 894	548	1 214.06
浙　江 Zhejiang	4 459	688	98.81
福　建 Fujian	2 623	832	148.02
山　东 Shandong	1 693	287	328.30 ①
广　东 Guangdong	289	64	207.16
广　西 Guangxi	80	23	0.81
海　南 Hainan	18	78	0.83

注：①表示单位为万载重吨。

Note: ①the unit is 10 thousand dead weight tonnage.

3-12 沿海地区海洋货物运输量和周转量
Volume of Maritime Goods Transported and Turnover by Coastal Regions

单位：万吨，亿吨·千米 (10 000 t, 100 million t-km)

地 区 Region	货运量 Volume of Goods Transported	沿 海 Coastal	远 洋 Oceangoing	货物周转量 Volume of Goods Turnover	沿 海 Coastal	远 洋 Oceangoing
全国总计 National Total	**225 870**	**155 049**	**70 821**	**66 845**	**18 405**	**48 440**
天 津 Tianjin	8 678	6 931	1 747	2 268	1 070	1 198
河 北 Hebei	3 048	2 508	540	863	413	450
辽 宁 Liaoning	13 379	7 323	6 056	7 837	428	7 409
上 海 Shanghai	37 514	23 969	13 545	13 925	3 724	10 201
江 苏 Jiangsu	23 350	15 838	7 512	6 416	1 763	4 653
浙 江 Zhejiang	49 655	46 145	3 510	7 009	4 917	2 092
福 建 Fujian	20 264	18 032	2 232	2 942	2 263	679
山 东 Shandong	8 601	7 770	831	997	684	313
广 东 Guangdong	32 764	16 276	16 488	5 452	2 100	3 352
广 西 Guangxi	5 214	4 832	382	716	705	11
海 南 Hainan	6 075	5 104	971	532	297	235
其 他* Others	17 328	321	17 007	17 887	38	17 849

注：其他数据为中国远洋运输（集团）总公司完成的远洋运输量。

Note: Other data are those of the freight volume of ocean transportation accomplished by the China National Ocean Shipping Corporation.

3-13 沿海地区海洋旅客运输量和周转量
Volume of Maritime Passenger Traffic and Turnover by Coastal Regions

单位：万人，亿人·千米 (10 000 persons, 100 million person-km)

地 区 Region	客运量 Passenger Traffic	沿 海 Coastal	远 洋 Oceangoing	旅客周转量 Passenger Turnover Volume	沿 海 Coastal	远 洋 Oceangoing
全国总计 National Total	**8 877**	**7 033**	**1 844**	**35.72**	**22.41**	**13.31**
天 津 Tianjin						
河 北 Hebei						
辽 宁 Liaoning	534	519	15	6.52	5.72	0.80
上 海 Shanghai	243	243		0.62	0.62	
江 苏 Jiangsu	49	27	22	1.83	0.05	1.78
浙 江 Zhejiang	2 288	2 288		4.22	4.22	
福 建 Fujian	1 400	1 346	54	2.42	1.83	0.59
山 东 Shandong	1 395	1 306	89	10.73	7.18	3.55
广 东 Guangdong	1 683	19	1 664	6.60	0.01	6.59
广 西 Guangxi	155	155		0.69	0.69	
海 南 Hainan	1 130	1 130		2.09	2.09	

3-14 沿海港口客货吞吐量
Volume of Passenger and Freight Handled at Coastal Seaports

单位：万吨，万人·次 (10 000 t, 10 000 person-times)

地　区 Region	货物吞吐量 Cargo Handled	#外　贸 Foreign Trade	旅客吞吐量 Passenger Leaving & Arriving	#离　港 Leaving
合　计 Total	**756 129**	**305 686**	**7 780**	**3 927**
天　津 Tianjin	50 063	26 738	33	17
河　北 Hebei	88 984	24 380	5	2
辽　宁 Liaoning	98 354	19 781	631	320
上　海 Shanghai	68 273	37 706	134	68
江　苏 Jiangsu	23 908	11 100	12	6
浙　江 Zhejiang	100 591	40 740	719	356
福　建 Fujian	45 475	18 564	1 008	505
山　东 Shandong	118 137	65 662	1 298	641
广　东 Guangdong	130 831	47 093	2 598	1 326
广　西 Guangxi	18 674	11 548	20	10
海　南 Hainan	12 839	2 374	1 322	676

3-15 沿海地区水路国际标准集装箱运量
Volume of International Standardized Containers Traffic by Coastal Regions

单位：万标准箱，万吨 (10 000 TEU, 10 000 t)

地 区 Region	2011		2012		2013	
	箱 数 Number of Containers	重 量 Weight	箱 数 Number of Containers	重 量 Weight	箱 数 Number of Containers	重 量 Weight
合 计 Total	**4 074**	**49 348**	**4 357**	**52 301**	**4 646**	**54 765**
天 津 Tianjin	8	85	6	62	9	168
河 北 Hebei	1	36	1	20	1	18
辽 宁 Liaoning	45	521	42	473	41	427
上 海 Shanghai	1 759	22 981	2 010	26 378	1 817	23 722
江 苏 Jiangsu	462	3 961	479	4 063	508	4 218
浙 江 Zhejiang	168	2 781	220	2 948	239	3 291
福 建 Fujian	268	4 673	302	4 373	345	5 419
山 东 Shandong	123	2 345	139	1 927	120	1 414
广 东 Guangdong	887	7 342	786	6 618	975	7 244
广 西 Guangxi	181	2 367	210	2 598	200	3 080
海 南 Hainan	172	2 256	163	2 839	111	1 874
其 他 other					280	3 890

3-16 沿海港口国际标准集装箱吞吐量
International Standardized Containers Handled at Coastal Seaports

单位：万标准箱，万吨 (10 000TEU, 10 000 t)

地 区 Region	2011		2012		2013	
	箱 数 Number of Containers	重 量 Weight	箱 数 Number of Containers	重 量 Weight	箱 数 Number of Containers	重 量 Weight
合 计 Total	**14 632**	**157 969**	**15 797**	**175 986**	**16 968**	**194 693**
天 津 Tianjin	1 159	11 929	1 230	13 442	1 301	15 216
河 北 Hebei	77	1 251	90	1 351	135	2 051
辽 宁 Liaoning	1 200	19 115	1 514	24 733	1 798	28 381
上 海 Shanghai	3 174	31 220	3 253	32 480	3 362	34 243
江 苏 Jiangsu	488	4 626	504	5 012	554	5 536
浙 江 Zhejiang	1 584	15 747	1 759	17 811	1 910	19 659
福 建 Fujian	970	12 099	1 073	13 308	1 169	14 890
山 东 Shandong	1 691	17 738	1 899	19 912	2 076	22 758
广 东 Guangdong	4 103	41 315	4 256	44 495	4 420	47 947
广 西 Guangxi	74	1 165	82	1 358	100	1 703
海 南 Hainan	112	1 764	137	2 084	142	2 310

3-17 沿海城市国内旅游人数
Domestic Visitors by Coastal Cities

单位：万人・次 (10 000 person-times)

城 市	City	2010	2011	2012
合 计	**Total**	**107 019**	**135 468**	**117 625**
天 津	**Tianjin**	6 118	10 605	
河 北	**Hebei**	3 968	4 790	5 977
唐 山	Tangshan	1 532	2 001	2 454
秦皇岛	Qinhuangdao	1 861	2 101	2 313
沧 州	Cangzhou	575	688	1 210
辽 宁	**Liaoning**	11 655	13 586	15 225
大 连	Dalian	3 777	4 261	4 687
丹 东	Dandong	2 248	2 646	2 990
锦 州	Jinzhou	1 442	1 704	1 926
营 口	Yingkou	1 074	1 279	1 446
盘 锦	Panjin	1 654	1 959	2 214
葫芦岛	Huludao	1 460	1 736	1 962
上 海	**Shanghai**	21 463	23 079	
江 苏	**Jiangsu**	4 255	5 090	5 839
南 通	Nantong	1 757	2 109	2 408
连云港	Lianyungang	1 393	1 656	1 894
盐 城	Yancheng	1 105	1 325	1 537
浙 江	**Zhejiang**	26 371	30 547	35 048
杭 州	Hangzhou	6 305	7 181	8 237
宁 波	Ningbo	4 624	5 181	5 748
温 州	Wenzhou	3 537	4 123	4 887
嘉 兴	Jiaxing	3 070	3 536	4 101
绍 兴	Shaoxing	3 436	4 128	4 866
舟 山	Zhoushan	2 113	2 433	2 740
台 州	Taizhou	3 286	3 965	4 469
福 建	**Fujian**	8 262	9 280	11 665
福 州	Fuzhou	2 275	2 680	3 107
厦 门	Xiamen	2 178	2 569	2 979
莆 田	Putian	775	1 165	1 112
泉 州	Quanzhou	1 351	1 083	2 170
漳 州	Zhangzhou	1 002	951	1 348
宁 德	Ningde	681	832	949

3-17 续表 continued

城　市 City		2010	2011	2012
山　东	**Shandong**	16 055	18 807	21 660
青　岛	Qingdao	4 397	4 956	5 591
东　营	Dongying	637	775	938
烟　台	Yantai	3 272	3 863	4 450
潍　坊	Weifang	2 946	3 603	4 221
威　海	Weihai	2 112	2 372	2 669
日　照	Rizhao	2 031	2 426	2 795
滨　州	Binzhou	661	812	996
广　东	**Guangdong**	13 612	15 538	17 411
广　州	Guangzhou	3 692	3 816	4 017
深　圳	Shenzhen	2 265	2 628	2 941
珠　海	Zhuhai	1 055	1 215	1 299
汕　头	Shantou	769	890	1 026
江　门	Jiangmen	872	1 013	1 120
湛　江	Zhanjiang	602	909	1 247
茂　名	Maoming	305	360	399
惠　州	Huizhou	913	1 014	1 122
汕　尾	Shanwei	327	439	518
阳　江	Yangjiang	305	456	572
东　莞	Dongguan	1 289	1 400	1 432
中　山	Zhongshan	540	605	744
潮　州	Chaozhou	317	361	438
揭　阳	Jieyang	361	432	536
广　西	**Guangxi**	1 958	2 347	2 811
北　海	Beihai	938	1 101	1 311
防城港	Fangchenggang	550	676	807
钦　州	Qinzhou	469	570	693
海　南	**Hainan**	1 564	1 799	1 989
海　口	Haikou	723	831	935
三　亚	Sanya	841	968	1 054

注：数据来源于《中国区域经济统计年鉴》（2013）。

Note：The data come from *China Statistical Yearbook For Regional Economy (2013)*.

3-18 主要沿海城市国际旅游（外汇）收入
(Foreign Exchange) Earnings from International Tourism in Major Coastal Cities

单位：万美元 (10 000 USD)

城　市	City	2011	2012	2013
天　津	Tianjin	175 553	222 641	259 128
秦皇岛	Qinhuangdao	13 770	19 454	25 573
大　连	Dalian	80 519	87 349	81 341
上　海	Shanghai	575 118	549 323	524 470
南　通	Nantong	39 916	42 995	11 196
连云港	Lianyungang	12 869	14 434	1 668
杭　州	Hangzhou	195 710	220 165	216 047
宁　波	Ningbo	65 472	73 428	79 656
温　州	Wenzhou	25 602	31 887	42 064
福　州	Fuzhou	102 854	110 817	128 932
厦　门	Xiamen	129 901	157 728	160 712
泉　州	Quanzhou	79 661	90 446	105 483
漳　州	Zhangzhou	18 781	22 878	21 855
青　岛	Qingdao	68 933	82 459	79 363
烟　台	Yantai	46 816	48 146	46 313
威　海	Weihai	21 855	25 283	23 851
广　州	Guangzhou	485 306	514 458	516 884
深　圳	Shenzhen	374 474	432 882	453 102
珠　海	Zhuhai	106 685	95 045	83 767
汕　头	Shantou	5 071	5 175	5 431
湛　江	Zhanjiang	3 630	4 816	5 845
中　山	Zhongshan	24 717	21 979	23 891
北　海	Beihai	2 574	3 429	4 308
海　口	Haikou	3 845	4 474	4 210
三　亚	Sanya	31 259	26 565	25 991

3-19 主要沿海城市接待入境旅游者人数
Number of Inbound Tourists Received by Major Coastal Cities

单位：人·次 (person-time)

城 市	City	2011	2012	2013
天 津	Tianjin	730 615	737 481	758 594
秦皇岛	Qinhuangdao	264 372	286 401	188 041
大 连	Dalian	1 170 035	1 284 176	734 605
上 海	Shanghai	6 686 144	6 512 347	6 140 911
南 通	Nantong	404 852	440 788	216 944
连云港	Lianyungang	132 289	144 684	24 228
杭 州	Hangzhou	3 063 140	3 311 225	1 060 360
宁 波	Ningbo	1 073 872	1 162 088	426 258
温 州	Wenzhou	470 504	575 397	290 335
福 州	Fuzhou	761 665	851 328	568 853
厦 门	Xiamen	1 799 205	2 124 163	1 072 263
泉 州	Quanzhou	846 389	951 424	622 476
漳 州	Zhangzhou	293 728	330 669	238 194
青 岛	Qingdao	1 156 391	1 270 076	782 632
烟 台	Yantai	548 533	530 184	328 872
威 海	Weihai	415 114	456 594	278 003
广 州	Guangzhou	7 786 900	7 866 031	7 681 966
深 圳	Shenzhen	11 045 500	12 064 451	12 148 917
珠 海	Zhuhai	3 208 300	2 975 791	2 632 331
汕 头	Shantou	140 700	147 800	155 222
湛 江	Zhanjiang	142 300	147 400	190 247
中 山	Zhongshan	608 000	558 573	537 743
北 海	Beihai	83 073	98 759	80 026
海 口	Haikou	146 808	179 741	156 632
三 亚	Sanya	528 942	481 437	481 851

3-20 主要沿海城市接待入境旅游者情况

Breakdown of Inbound Tourists Received by Major Coastal Cities

单位：人·次，人·天 (person-time, night)

城市 City	外国人 Foreigners		香港同胞 Hong Kong	
	人次数 Arrivals	人天数 Nights	人次数 Arrivals	人天数 Nights
天　津 Tianjin	660 355	9 744 194	46 862	1 017 779
秦皇岛 Qinhuangdao	176 171	675 478	5 446	19 444
大　连 Dalian	567 327	1 701 981	83 354	270 511
上　海 Shanghai	5 110 746	16 574 040	418 957	1 291 313
南　通 Nantong	196 554	496 741	4 133	9 750
连云港 Lianyungang	20 486	58 153	424	883
杭　州 Hangzhou	762 818	1 790 238	125 291	276 370
宁　波 Ningbo	320 905	653 537	42 229	85 714
温　州 Wenzhou	218 701	630 157	27 849	57 188
福　州 Fuzhou	329 534	2 380 808	73 002	491 657
厦　门 Xiamen	512 608	2 560 574	129 507	596 070
泉　州 Quanzhou	118 034	688 194	347 160	1 897 941
漳　州 Zhangzhou	58 647	239 364	67 259	254 019
青　岛 Qingdao	539 251	2 017 339	123 661	469 170
烟　台 Yantai	257 567	1 137 585	22 189	104 896
威　海 Weihai	261 201	815 623	2 287	5 991
广　州 Guangzhou	2 789 884	7 293 411	3 845 866	8 123 626
深　圳 Shenzhen	1 668 230	3 719 798	9 995 083	20 011 458
珠　海 Zhuhai	483 617	1 051 018	972 868	1 724 961
汕　头 Shantou	96 022	191 100	49 000	91 100
湛　江 Zhanjiang	86 747	185 100	81 300	177 400
中　山 Zhongshan	103 819	401 735	317 632	852 311
北　海 Beihai	42 335	74 386	26 025	43 335
海　口 Haikou	67 571	133 785	25 560	41 871
三　亚 Sanya	369 895	1 265 021	56 100	113 502

3-20 续表 continued

城 市 City	澳门同胞 Macao		台湾同胞 Taiwan Province	
	人次数 Arrivals	人天数 Nights	人次数 Arrivals	人天数 Nights
天 津 Tianjin	3 114	67 873	48 263	1 002 768
秦皇岛 Qinhuangdao	439	1 577	5 985	20 476
大 连 Dalian	1 472	4 710	82 452	263 846
上 海 Shanghai	16 501	63 491	594 707	2 236 500
南 通 Nantong	116	245	16 141	53 529
连云港 Lianyungang	13	26	3 305	13 003
杭 州 Hangzhou	14 260	23 547	157 991	400 160
宁 波 Ningbo	2 302	3 457	60 822	182 227
温 州 Wenzhou	10 424	10 709	33 361	124 459
福 州 Fuzhou	5 234	32 126	161 083	637 547
厦 门 Xiamen	5 425	23 913	424 723	1 914 742
泉 州 Quanzhou	35 632	169 038	121 650	580 560
漳 州 Zhangzhou	6 609	19 481	105 679	302 806
青 岛 Qingdao	28 957	82 924	90 763	330 315
烟 台 Yantai	9 761	31 427	39 355	147 318
威 海 Weihai	582	1 548	13 933	45 166
广 州 Guangzhou	495 293	1 042 141	550 923	1 400 680
深 圳 Shenzhen	53 704	99 177	431 900	1 016 047
珠 海 Zhuhai	668 223	1 365 413	507 623	1 172 832
汕 头 Shantou	700	1 400	9 500	18 700
湛 江 Zhanjiang	6 500	14 900	15 700	31 600
中 山 Zhongshan	68 253	156 780	48 039	162 253
北 海 Beihai	3 774	6 397	7 892	13 785
海 口 Haikou	1 296	2 028	62 205	78 428
三 亚 Sanya	3 679	6 269	52 177	92 769

主要统计指标解释

1. 海洋捕捞产量 凡是从海洋里捕捞的天然生长的水产品产量为捕捞产量。

2. 海水养殖产量 凡是从人工投放苗种或天然纳苗并进行人工饲养管理的海水养殖水域中捕捞的水产品产量为海水养殖产量。

3. 远洋捕捞产量 由各远洋渔业企业和各生产单位按我国远洋渔业项目管理办法组织的远洋渔船（队）在非我国管辖海域（外国专属经济区水域或公海）捕捞的水产品产量。中外合资、合作渔船捕捞的水产品只统计按协议应属于中方所有的部分。

4. 原油产量 是按净原油量来计算的，能直接用于销售和生产自用的原油量。目前海洋石油系统原油产量计算方法采用倒算法。

原油产量=销售量+期末库存量-期初库存量+海上平台及陆地终端处理厂自用量。

5. 天然气产量 指进入集输管网的销售量和就地利用的全部气量。

天然气产量=外输（销）量+企业自用量

6. 海洋原油出口量 指销往国外的产品数量。

7. 海洋原油出口创汇额 指产品销往国外的归中方所有的全部外汇收入。以美元或万美元表示。

8. 造船综合吨 等于以计量单位载重吨和满载排水量吨的民用船舶的吨位数之和。

9. 货运量 指经船舶实际运送的货物重量。

10. 货物周转量 指实际运送的货物与其运送距离的乘积。

11. 集装箱运量 既包含货重，也包含箱重。箱重系指承运租用的空箱重量凡有运费收入的空箱，其重量应统计为运量，按空箱 1 吨为货运量 1 吨计算；若无收入，所承运的空箱一律不作运量统计。

12. 旅客周转量 指实际运送的旅客人数与其运送距离的乘积。

13. 国际旅游外汇收入 入境游客在中国（大陆）境内旅行、游览过程中用于交通、参观游览、住宿、餐饮、购物、娱乐等的全部花费。

14. 接待人次数 指报告期内我国接待游客人数。游客按出游地分为入境游客和国内游客，按出游时间分为旅游者（过夜游客）和一日游游客（不过夜游客）。

15. 接待人天数 指过夜旅游者的停留天数。

16. 外国人 指外国国籍的人，加入外籍的中国血统华人也计入外国人。

17. 港澳台同胞 指居住在我国香港特别行政区、澳门特别行政区和台湾省的中国同胞。

Explanatory Notes on Main Statistical Indicators

1. Marine Catches refers to the output of the naturally growing aquatic products caught from the sea.

2. Mariculture Production refers to the output of aquatic products whose young are artificially released or naturally collected, and raised and managed artificially, and which are caught from the waters of mariculture.

3. Deep-Sea Fishing Production refers to the output of aquatic products caught in the non-Chinese jurisdictional sea areas (foreign EEE or high sea) by the distant fishing vessels (fleet) organized by various distant fishing businesses and production units according to the management measures of the China distant fishing projects. The aquatic products caught by the Chinese-foreign joint ventures' and cooperative fishing vessels are counted only for the part owned by the Chinese side according to the agreement.

4. Output of Crude Oil is calculated on the basis of the net amount of crude oil, i.e., the amount of crude oil that may be directly used for sale and for the production itself.

Output of crude oil = Volume of sales + Reserves at the end of the period - Reserves at the beginning of the period +Amount for self-use on the platforms and in the terminal processing plants on land.

5. Output of Natural Gas refers to the total gas volume of the sales volume entering the oil collecting and transport pipeline network and that used locally.

Output of natural gas = Volume of sales or transport to other areas + Volume used by the enterprise itself

6. Export Volume of Offshore Crude Oil refers to the amount of products for sale abroad.

7. Foreign Exchange Earnings of Offshore Crude Oil refer to the total foreign exchange income from oil (gas) products for sale abroad which is owned by the Chinese side.

8. Comprehensive Tonnages of Shipbuilding refers to the sum of tonnage of civilian vessels with the deadweight capacity and full-load displacement as measured.

9. Freight Traffic refers to the weight of cargoes actually transported by vessels.

10. Cargoes Turnover Volume refers to the product of the actually transported cargoes and the transport distance.

11. Freight Volume of Containers includes the weight of both cargoes and container boxes. The weight of container boxes refers to the weight of empty containers rented for transport or having freight income and should be included in the freight volume, one ton of empty boxes equalling to one ton of freight volume. The empty boxes which have no income for transportation are not included in the freight volume.

12. Passenger Turnover Volume refers to the product of the number of passengers actually transported and the shipping distance.

13. International Tourism (Foreign Exchange) Receipts refer to the total expenditure made by inbound visitors within the territory of China (the mainland) in their course of travel on transport, tours and sightseeing, lodging, food and beverage, shopping, entertainment, etc.

14. Number of Person-Times Received refers to the number of visitors received by China in the period reported. Visitors are divided into inbound visitors and domestic visitors by origin of the travel, and tourists (overnight visitors) and same-day visitors by their length of stay.

15. Number of the Days of Stay refers to the number of the days of stay of tourists.

16. Foreigners refer to the people with foreign nationality, including foreign nationals of Chinese descent.

17. Compatriots from Hong Kong, Macao and Taiwan Province refer to the Chinese compatriots living in the Hong Kong Special Administrative Region, the Macao Special Administrative Region and Taiwan Province.

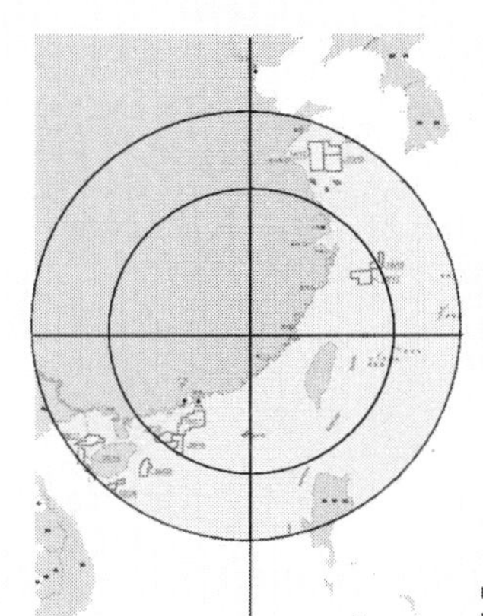

4

主要海洋产业生产能力

Production Capacity of Major Marine Industries

4-1　沿海地区渔港情况
Fishing Ports in the Coastal Regions

单位：个　　(unit)

地　区 Region		渔港合计 The Total of the Fishing Ports	中心渔港 Central Fishing Port	一级渔港 Grade 1 Fishing Port
合　计	**Total**	**128**	**58**	**70**
天　津	Tianjin			
河　北	Hebei	7	3	4
辽　宁	Liaoning	10	3	7
上　海	Shanghai			
江　苏	Jiangsu	11	6	5
浙　江	Zhejiang	22	9	13
福　建	Fujian	21	9	12
山　东	Shandong	20	10	10
广　东	Guangdong	19	8	11
广　西	Guangxi	8	4	4
海　南	Hainan	10	6	4

资料来源：《2014中国渔业统计年鉴》。
Data Source: *China Fishery Statistical Yearbook 2014*.

4-2 沿海地区海水养殖面积
Mariculture Area by Coastal Regions

单位：公顷 (hm²)

地 区 Region	海水养殖面积 Mariculture Area
合 计 Total	**2 315 569**
天 津 Tianjin	3 169
河 北 Hebei	117 928
辽 宁 Liaoning	942 050
上 海 Shanghai	
江 苏 Jiangsu	193 807
浙 江 Zhejiang	89 358
福 建 Fujian	154 453
山 东 Shandong	546 814
广 东 Guangdong	197 198
广 西 Guangxi	54 001
海 南 Hainan	16 791

4-3 海洋油气勘探情况
Work Volume of Offshore Oil and Gas

地 区 Region	地震测线 Seismic Line		钻井（口） Drilling (well)	
	二维 （千米） Two Dimensions (km)	三维 （平方千米） Three Dimensions (km^2)	预探井 Wildcat Wells	评价井 Appraisal Wells
合 计 Total	**28 069**	**25 712**	**72**	**94**
天 津 Tianjin		6 771	20	36
其中：合作 Including: Cooperative		261		
河 北 Hebei	290	920	13	18
辽 宁 Liaoning			1	2
上 海 Shanghai	3 281	4 566	3	2
山 东 Shandong			5	10
广 东 Guangdong	24 498	13 455	30	26
其中：合作 Including: Cooperative		5 732	3	

4-4 海洋油气生产井情况
Survey of Offshore Oil and Gas Production Wells

单位：口 (well)

地 区 Region	合 计 Total	采油井 Oil Wells	采气井 Gas Wells	注水井 Injection Wells	其他井 Others
合 计 Total	**6 043**	**4 472**	**290**	**1 270**	**11**
天 津 Tianjin	3 376	2 489	146	741	
河 北 Hebei	1 065	795	5	265	
辽 宁 Liaoning	241	184	6	40	11
上 海 Shanghai	47	13	34		
山 东 Shandong	653	446	8	199	
广 东 Guangdong	661	545	91	25	

4-5 沿海地区盐田面积和海盐生产能力
Salt Pan Area and Sea Salt Production Capacity by Coastal Regions

地 区 Region	盐田总面积（公顷） Total Area of Salt Pan (hm^2)		生产面积（公顷） Production Area (hm^2)		年末海盐生产能力（万吨） Year-End Capacity of Sea Salt Production (10 000 t)	
	2012	2013	2012	2013	2012	2013
合 计 Total	**441 482**	**418 646**	**323 196**	**303 585**	**3 641.41**	**3 363.79**
天 津 Tianjin	27 258	27 277	26 470	26 499	166.56	160.00
河 北 Hebei	84 350	77 335	63 951	59 093	465.10	436.35
辽 宁 Liaoning	34 632	33 938	29 091	28 532	179.17	178.67
江 苏 Jiangsu	60 209	54 976	19 179	15 603	87.81	72.00
浙 江 Zhejiang	2 754	2 296	2 247	1 852	13.16	11.66
福 建 Fujian	6 316	5 115	4 884	4 353	37.94	33.00
山 东 Shandong	209 753	202 365	167 316	157 908	2 648.24	2 431.63
广 东 Guangdong	10 052	10 052	6 014	6 014	18.62	16.49
广 西 Guangxi	2 504	1 686	1 236	948	6.49	5.99
海 南 Hainan	3 654	3 606	2 808	2 783	18.32	18.00

4-6　沿海地区风能发电能力
Wind Power Production by Coastal Regions

单位：万千瓦　　　　(10 000 kW)

地　区 Region	风能年发电能力 Annual Wind Power Generation Capacity	
	2012	2013
合　计 Total	**1 352.94**	**1 588.99**
天　津 Tianjin	27.80	30.50
辽　宁 Liaoning	142.94	170.69
河　北 Hebei	22.05	22.05
上　海 Shanghai	35.20	37.00
江　苏 Jiangsu	230.55	268.46
浙　江 Zhejiang	46.67	59.53
福　建 Fujian	128.57	154.42
山　东 Shandong	534.06	628.69
广　东 Guangdong	154.38	186.93
广　西 Guangxi	0.25	0.25
海　南 Hainan	30.47	30.47

注：本表数据（除海南省）为沿海城市合计数，海南省为沿海地带合计数。
Note: The data in this table (except that for Hainan Province) are for the total number of coastal cities and these for Hainan Province are for the total number of coastal zones.

4-7 主要潮汐电站分布情况
Distribution of Major Tidal Power Stations

电站名称 Name	运行情况 Status of Operation	装机容量（千瓦） Capacity (kW)
江厦潮汐试验电站	1970年开始建造，1980年投入使用，运行至今	3 900
Jiangxia Experimental Tidal Power Station	It began construction in 1970, was put into use in 1980, and has been in operation so far	
海山潮汐电站	1972年开始建造，1975年投入使用，停止运行	
Haishan Tidal Power Station	It began construction in 1972, was put into use in 1975, and stopped power generation	
岳浦潮汐电站	1970年开始建造，1978年停止运行	
Yuepu Tidal Power Station	It began construction in 1970, and stopped power generation in 1978	
白沙口潮汐电站	1970年开始建造，1978年投入使用，2010年停止运行	
Baishakou Tidal Power Station	It began construction in 1970, was put into use in 1978, and stopped power generation in 2010	

4-8 主要海上活动船舶
Major Vessels Operating on the Sea

类 别 Type	艘数（艘） Number of Vessels (unit)	总吨（万吨位） grt (10 000 t)	净载重量（万吨） Net Weight Tonnage (10 000 t)	载客量（客位） Passenger Spaces (seat)	总功率（千瓦） Total Power (kW)
一、海洋生产用船 Vessels for Marine Production					
海洋渔业船舶 Marine Fishing Vessels	283 321	813.32			16 396 732
远洋渔船 Ocean-going Fishing Vessels	2 159				1 583 513
海洋油气船舶 Offshore Oil and Gas Vessels	177	106.00	53.00	10 425	920 950
钻井平台 Drilling Vessels	43	41.00	12.00	5 820	290 807
物探船 Physical Exploration Vessels	9	3.00	1.00	469	42 601
其 他 Others	125	62.00	40.00	4 136	587 542
海洋运输船舶 Maritime Transport Vessels	12 764	9 111.39	13 808.47	216 035	33 820 735
二、海洋科研用船 Vessels for Marine Scientific Research					
海洋地质勘探船 Marine Geological Survey Vessels	6	1.19	0.37	425	23 371
海洋调查船 Marine Research Vessels					
中国科学院 Chinese Academy of Sciences	6	1.49	0.20	264	30 220
国家海洋局 State Oceanic Administration	4	2.60	1.35	318	25 012
三、海洋执法用船 Vessels for Marine Law Enforcement					
海监船* Marine Surveillance Vessels	562				
渔政执法船 Fishing law Enforcement	557	5.66			303 320

注：*包含船、艇数据。

Note: * include the data of ships and boats.

4-9 沿海规模以上港口生产用码头泊位
Berths for Productive Use at the coastal Seaports above Designed Size

单位：米, 个 (m, unit)

港口 Seaport		码头长度 Length of Quay Line	泊位个数 Number of Berths	#万吨级 10 000 Tonnage Class
合　计	**Total**	**669 379**	**4 841**	**1 524**
丹　东	Dandong	6 407	38	21
大　连	Dalian	38 149	212	96
营　口	Yingkou	16 363	76	50
锦　州	Jinzhou	6 119	23	21
秦皇岛	Qinhuangdao	14 750	66	42
黄　骅	Huanghua	6 642	29	23
唐　山	Tangshan	20 360	77	74
天　津	Tianjin	34 408	149	102
烟　台	Yantai	17 911	88	59
威　海	Weihai	3 949	20	12
青　岛	Qingdao	20 944	82	66
日　照	Rizhao	12 707	50	44
上　海	Shanghai	74 487	608	156
连云港	Lianyungang	11 715	51	44
嘉　兴	Jiaxing	7 975	54	27
宁波-舟山	Ningbo-Zhoushan	78 371	613	146
台　州	Taizhou	11 325	174	7
温　州	Wenzhou	14 883	209	16

4-9 续表 continued

港口 Seaport		码头长度 Length of Quay Line	泊位个数 Number of Berths	#万吨级 10 000 Tonnage Class
福州	Fuzhou	22 824	181	48
莆田	Putian	5 002	46	8
泉州	Quanzhou	15 276	104	24
厦门	Xiamen	25 596	146	65
汕头	Shantou	9 627	87	19
汕尾	Shanwei	1 286	14	1
惠州	Huizhou	9 048	39	18
深圳	Shenzhen	29 384	147	66
虎门	Humen	13 804	108	23
广州	Guangzhou	45 477	498	66
中山	Zhongshan	3 837	64	0
珠海	Zhuhai	17 047	148	27
江门	Jiangmen	9 803	149	2
阳江	Yangjiang	2 232	10	9
茂名	Maoming	2 428	18	9
湛江	Zhanjiang	15 542	146	30
北部湾港	Beibuwan	30 902	241	66
海口	Haikou	4 372	31	10
洋浦	Yangpu	6 273	33	18
八所	Basuo	2 154	12	9

4-10 沿海地区星级饭店基本情况
Star Grade Hotels and Occupancies by Coastal Regions

地 区 Region	饭店数（座） Number of Hotels (unit)	客房数（间） Number of Rooms (unit)	床位数（张） Number of Beds (unit)	客房出租率（%） Room Occupancy (%)
合 计 Total	**5 383**	**771 791**	**1 318 765**	**55.10**
天 津 Tianjin	93	17 078	27 812	49.79
河 北 Hebei	409	52 499	98 694	48.31
辽 宁 Liaoning	421	54 748	94 421	54.11
上 海 Shanghai	256	61 571	95 834	59.25
江 苏 Jiangsu	735	93 820	157 653	57.89
浙 江 Zhejiang	828	113 787	193 136	54.83
福 建 Fujian	401	58 031	97 674	57.79
山 东 Shandong	792	95 127	168 939	55.73
广 东 Guangdong	917	146 820	243 791	56.71
广 西 Guangxi	381	50 729	91 906	55.56
海 南 Hainan	150	27 581	48 905	56.13

注：本表数据为主要沿海城市合计数。
Note: The data in this table are for the total number of major coastal cities.

4-11 沿海地区旅行社数
Number of Travel Agencies by Coastal Regions

单位：家 (unit)

地 区 Region		旅行社总数 Number of Travel Agencies
合 计	**Total**	**13 325**
天 津	Tianjin	383
河 北	Hebei	1 271
辽 宁	Liaoning	1 165
上 海	Shanghai	1 139
江 苏	Jiangsu	2 073
浙 江	Zhejiang	1 988
福 建	Fujian	784
山 东	Shandong	2 001
广 东	Guangdong	1 656
广 西	Guangxi	513
海 南	Hainan	352

主要统计指标解释

1. 海水养殖面积　是指利用海上、滩涂、陆基进行鱼、甲壳类（虾、蟹）、贝、藻等海水经济动植物的人工养殖的水面面积。在报告期内无论是否全部收获或尚未收获其产品，均应统计在海水养殖面积中。但有些滩涂、水面不投放苗种或投放少量苗种，只进行一般管理的，不统计为养殖面积。

2. 盐田总面积　指盐田占有的全部面积。包括储卤、蒸发、保卤、结晶面积、滩内的沟、壕、池、埝、滩坨等面积及滩外的沟、壕、公路及杂地面积。

3. 生产面积　指直接提供给海盐生产的面积，包括结晶面积、蒸发面积、保卤面积，滩内的沟、壕、池、埝面积及滩坨面积。

4. 年末海盐生产能力　指年末企业生产原盐的全部设备的综合平衡能力。海盐生产露天作业，受天气影响，因而计算生产能力时，成熟滩田按 10 年实际平均单位生产面积产量乘以本年成熟滩田生产面积而得，新滩田按设计能力及滩田成熟程度可能达到的产量计算。

5. 海洋渔业船舶　是指配置机器作为动力的从事海洋渔业生产和辅助渔业生产的船舶。

6. 远洋渔船　按我国远洋渔业项目管理办法在非我国管辖海域（外国专属经济区水域或公海）进行常年或季节性生产的渔船。

7. 泊位个数　是指设有系靠船舶设施，在同一时间内可供靠泊最大吨级船舶的艘数。即可靠泊一艘船舶，则计为一个泊位，余类推。泊位分码头泊位和浮筒泊位。

8. 客房数　指饭店实际可用于接待旅游者的房间数。

9. 床位数　指饭店实际可用于接待旅游者的床位数。

Explanatory Notes on Main Statistical Indicators

1. Mariculture Area refers to the area of the water surface where seawater economic animals and plants, such as crustacean (shrimp, crab), shellfish and algae, are cultivated at sea, on tidal flat and land. Whether or not all the products in the area have been harvested or the products have not been harvested yet in the period covered by the report, the area is included in the Mariculture Area. But some tidal flats and water surfaces where none or a small amount of the young have been released and only general management is carried out are not included in the Mariculture Area.

2. Total Area of Salt Pans refers to the total area covered by salt pans, including the area for brine storage, evaporation, brine preservation, and crystallization, the area of ditches, moats, pondsand banks within the beach as well as beach mounds, and ditches, moats, highway beyond the beach as well as the area of miscellaneous lands.

3. Area of Salt Pan Production refers to the area directly provided for sea-salt productions, including the area for crystallization, evaporation and brine preservation, the area of ditches, moats, ponds and banks within the beach as well as the area of beach mounds.

4. Year-End Capacity of Crude Salt Production refers to the integrated and balanced capacity of all equipment of the enterprise used for crude salt production at the end of the year. As sea salt production is an open-air operation, which is subject to the effect of weather, the production capacity of a matured salt pan is calculated at the productions of the actual average unit production area in ten years times the production area of the matured salt pan in the current year. The production capacity of new salt pans is calculated at the production that may be reached in the light of the designed capacity and the level of maturity of the salt pan.

5. Marine Fishing Vessels refer to the vessels equipped with machines as motive power and going for marine fishery production and auxiliary fishery production.

6. Deep Sea Fishing Vessels refer to the fishing vessels which carry out production all the year round or seasonally in the non-Chinese jurisdictional, sea areas (foreign EEZ or high sea) according to the China Deep-Sea Fishing Projects Management Measures.

7. Number of Berths refers to the spaces equipped with facilities for docking ships and the number of ships of the maximum tonnage that may dock or anchor in them. A space for a ship to dock is counted as one berth and the rest are reasoned out by analogy. Berths are divided into wharf berths and buoy berths.

8. Number of Rooms refers to the number of guest rooms actually used by the hotels receiving tourists.

9. Number of Beds refers to the number of beds actually used by the hotels receiving tourists.

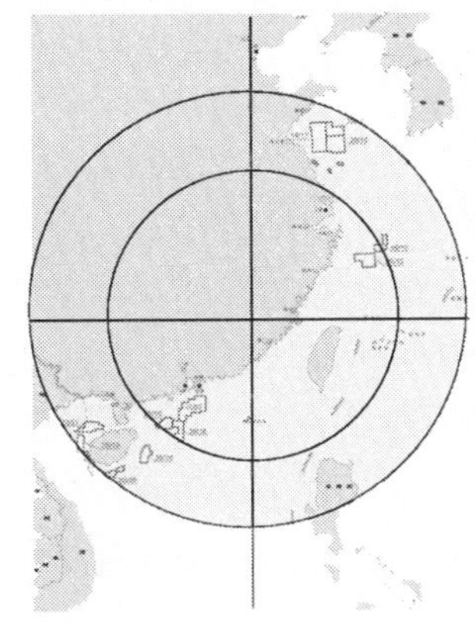

5

涉 海 就 业

Ocean-Related Employment

5-1 全国涉海就业人员情况
National Ocean-Related Employed Personnel by Region

单位：万人　　　　(10 000 persons)

地　区 Region	2001	2012①	2013①
合　计 Total	**2 107.6**	**3 468.8**	**3 514.3**
天　津 Tianjin	106.4	175.1	177.4
河　北 Hebei	58.0	95.5	96.7
辽　宁 Liaoning	196.0	322.6	326.8
上　海 Shanghai	127.5	209.8	212.6
江　苏 Jiangsu	116.9	192.4	194.9
浙　江 Zhejiang	256.4	422.0	427.5
福　建 Fujian	259.7	427.4	433.0
山　东 Shandong	319.9	526.5	533.4
广　东 Guangdong	505.3	831.6	842.6
广　西 Guangxi	68.9	113.4	114.9
海　南 Hainan	80.6	132.7	134.4
其　他② Others	12.0	19.7	20.0

注：① 2012年和2013年为推算数据；② 其他为非沿海地区涉海就业人员数。

Note: ① The data for 2012 and 2013 are the estimated ones;

② Others refer to the number of ocean-related employed persons in the non-coastal regions.

5-2 全国主要海洋产业就业人员情况
Employed Personnel in the Major Marine Industries Throughout the Country

单位：万人 (10 000 persons)

海洋产业 Marine Industry	2001	2012	2013
合计 Total	**719.1**	**1 183.5**	**1 199.1**
海洋渔业及相关产业 Marine Fishery and the Related Industries	348.3	573.2	580.8
海洋石油和天然气业 Offshore Oil and Natural Gas Industry	12.4	20.4	20.7
海滨砂矿业 Beach Placer Mining Industry	1.0	1.6	1.7
海洋盐业 Sea Salt Industry	15.0	24.7	25.0
海洋化工业 Marine Chemical Industry	16.1	26.5	26.8
海洋生物医药业 Marine Biomedicine Industry	0.6	1.0	1.0
海洋电力和海水利用业 Marine Electric Power and Seawater Utilization Industry	0.7	1.2	1.2
海洋船舶工业 Marine Shipbuilding Industry	20.6	33.9	34.3
海洋工程建筑业 Marine Engineering Architecture Industry	38.8	63.9	64.7
海洋交通运输业 Maritime Communications and Transportation Industry	50.8	83.6	84.7
滨海旅游业 Coastal Tourism	78.3	128.9	130.6
其他海洋产业 Other Marine Industries	136.5	224.7	227.6

注：2012年和2013年为推算数据。

Note: The data for 2012 and 2013 are the estimated ones.

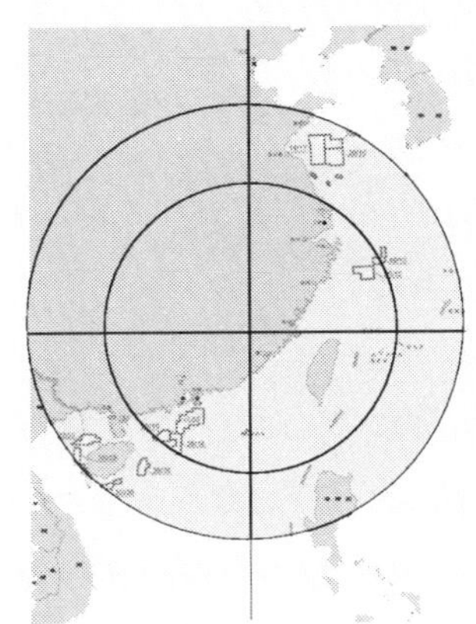

6

海洋科学技术

Marine Science and Technology

6-1 分行业海洋科研机构及人员情况
Marine Scientific Research Institutions and Personnel by Industry

行　业 Industry	机构数（个） Number of Institutions (unit)	从业人员（人） Employed Population (person)
合　计 **Total**	**175**	**38 754**
海洋基础科学研究 **Marine Basic Scientific Research**	**98**	**18 171**
海洋自然科学 Marine Natural Science	56	13 917
海洋社会科学 Marine Social Science	3	780
海洋农业科学 Marine Agricultural Science	37	3 388
海洋生物医药 Marine Biomedicine	2	86
海洋工程技术研究 **Marine Engineering Technology Research**	**65**	**18 894**
海洋化学工程技术 Marine Chemical Engineeing Technology	12	6 399
海洋生物工程技术 Marine Bioengineering Technology	2	257
海洋交通运输工程技术 Marine Communications and Transport Technology	14	3 932

6-1 续表 continued

行 业 Industry	机构数（个） Number of Institutions (unit)	从业人员（人） Employed Population (person)
海洋能源开发技术 Marine Energies Development Technology	4	3 165
海洋环境工程技术 Marine Environmental Engineering Technology	11	1 047
河口水利工程技术 Eustuarine Water Conservancy Engineering Technology	18	3 265
其他海洋工程技术 Other Marine Engineering Technology	4	829
海洋信息服务业 **Marine Information Service**	**9**	**1 023**
其他海洋信息服务 Other Marine Information Services	9	1 023
海洋技术服务业 **Marine Technological Service Industry**	**3**	**666**
其他海洋专业技术服务 Other Marine Professional and Technological Services	2	122
海洋工程管理服务 Marine Engineering Management Service	1	544

注：机构为县级以上科研机构；行业分类参照《海洋及相关产业分类》标准。

Note: The institutions are the scientific research institutions above the county level;The classification of industries follows the standard *Classification of Marine Industries and the Related Industries* .

6-2 分行业海洋科研机构科技活动人员学历构成
Educational Background Composition of the Personnel Engaged in Scientific and Technological Activities in Marine Scientific Research Institutions by Industry

单位：人 (person)

行 业 Industry	科技活动人员 Personnel Engaged in Scientific Activities	博士 Doctor	硕士 Master	大学生 University Graduate	大专生 College Graduate
合 计 Total	**32 349**	**7 499**	**9 352**	**10 067**	**3 045**
海洋基础科学研究 Marine Basic Scientific Research	**15 907**	**4 984**	**4 261**	**4 230**	**1 419**
海洋自然科学 Marine Natural Science	12 312	4 474	3 241	2 860	987
海洋社会科学 Marine Social Science	745	115	284	253	68
海洋农业科学 Marine Agricultural Science	2 781	394	727	1 076	355
海洋生物医药 Marine Biomedicine	69	1	9	41	9
海洋工程技术研究 Marine Engineering Technology Research	**14 865**	**2 398**	**4 520**	**5 145**	**1 489**
海洋化学工程技术 Marine Chemical Engineeing Technology	5 399	761	1 247	1 669	747
海洋生物工程技术 Marine Bioengineering Technology	221	16	90	74	25

6-2 续表 continued

行 业 Industry	科技活动人员 Personnel Engaged in Scientific Activities	博士 Doctor	硕士 Master	大学生 University Graduate	大专生 College Graduate
海洋交通运输工程技术 Marine Communications and Transport Technology	2 460	124	828	1 179	205
海洋能源开发技术 Marine Energies Development Technology	2 611	867	894	600	179
海洋环境工程技术 Marine Environmental Engineering Technology	914	58	297	476	56
河口水利工程技术 Eustuarine Water Conservancy Engineering Technology	2 732	493	899	995	256
其他海洋工程技术 Other Marine Engineering Technology	528	79	265	152	21
海洋信息服务业 Marine Information Service	**924**	**74**	**334**	**366**	**104**
其他海洋信息服务 Other Marine Information Services	924	74	334	366	104
海洋技术服务业 Marine Technological Service Industry	**653**	**43**	**237**	**326**	**33**
其他海洋专业技术服务 Other Marine Professional and Technological Services	109	5	14	67	16
海洋工程管理服务 Marine Engineering Management Service	544	38	223	259	17

6-3 分行业海洋科研机构科技活动人员职称构成
Professional Title Composition of Personnel Engaged in Scientific and Technological Activities in the Marine Scientific Research Institutions by Industry

单位：人 (person)

行 业 Industry	科技活动人员 Personnel Engaged in Scientific Activities	高级职称 Senior	中级职称 Intermediate	初级职称 Primary
合 计 Total	**32 349**	**12 380**	**10 928**	**4 998**
海洋基础科学研究 Marine Basic Scientific Research	**15 907**	**6 326**	**5 435**	**2 522**
海洋自然科学 Marine Natural Science	12 312	5 042	4 251	1 850
海洋社会科学 Marine Social Science	745	345	182	61
海洋农业科学 Marine Agricultural Science	2 781	929	983	571
海洋生物医药 Marine Biomedicine	69	10	19	40
海洋工程技术研究 Marine Engineering Technology Research	**14 865**	**5 487**	**4 903**	**2 152**
海洋化学工程技术 Marine Chemical Engineering Technology	5 399	1 992	1 658	525
海洋生物工程技术 Marine Bioengineering Technology	221	56	79	64

6-3 续表 continued

行 业 Industry	科技活动人员 Personnel Engaged in Scientific Activities	高级职称 Senior	中级职称 Intermediate	初级职称 Primary
海洋交通运输工程技术 Marine Communications and Transport Technology	2 460	744	777	602
海洋能源开发技术 Marine Energies Development Technology	2 611	927	989	424
海洋环境工程技术 Marine Environmental Engineering Technology	914	330	352	139
河口水利工程技术 Eustuarine Water Conservancy Engineering Technology	2 732	1 216	839	338
其他海洋工程技术 Other Marine Engineering Technology	528	222	209	60
海洋信息服务业 Marine Information Service	**924**	**324**	**322**	**221**
其他海洋信息服务 Other Marine Information Services	924	324	322	221
海洋技术服务业 Marine Technological Service Industry	**653**	**243**	**268**	**103**
其他海洋专业技术服务 Other Marine Professional and Technological Services	109	50	37	16
海洋工程管理服务 Marine Engineering Management Service	544	193	231	87

6-4 分行业海洋科研机构经费收入
Research Funds Receipts of the Marine Scientific Research Institutions by Industry

单位：万元 (10 000 yuan)

行 业 Industry	经费收入总额 Fund Total	经常费 Routine Fund	科技活动借贷款 Loans in the Scientific and Technological Activities	基本建设中政府投资 Government Investment in the Capital Construction
合 计 **Total**	**26 556 354**	**24 881 708**	**108 900**	**1 565 746**
海洋基础科学研究 **Marine Basic Scientific Research**	**12 753 101**	**11 530 232**	**0**	**1 222 869**
海洋自然科学 Marine Natural Science	10 920 050	9 853 254	0	1 066 796
海洋社会科学 Marine Social Science	437 074	380 273	0	56 801
海洋农业科学 Marine Agricultural Science	1 374 680	1 275 408	0	99 272
海洋生物医药 Marine Biomedicine	21 297	21 297	0	0
海洋工程技术研究 **Marine Technological Service Industry**	**12 927 048**	**12 487 788**	**108 900**	**330 360**
海洋化学工程技术 Marine Chemical Engineeing Technology	3 669 537	3 621 087	3 000	45 450
海洋生物工程技术 Marine Bioengineering Technology	204 413	204 413	0	0

6-4 续表 continued

行 业 Industry	经费收入总额 Fund Total	经常费 Routine Fund	科技活动借贷款 Loans in the Scientific and Technological Activities	基本建设中政府投资 Government Investment in the Capital Construction
海洋交通运输工程技术 Marine Communications and Transport Technology	2 843 271	2 561 301	105 900	176 070
海洋能源开发技术 Marine Energies Development Technology	3 286 030	3 286 030	0	0
海洋环境工程技术 Marine Environmental Engineering Technology	615 454	558 710	0	56 744
河口水利工程技术 Eustuarine Water Conservancy Engineering Technology	1 655 792	1 603 696	0	52 096
其他海洋工程技术 Other Marine Engineering Technology	652 551	652 551	0	0
海洋信息服务业 Marine Information Service	**564 820**	**552 303**	**0**	**12 517**
其他海洋信息服务 Other Marine Information Services	564 820	552 303	0	12 517
海洋技术服务业 Marine Technological Service Industry	**311 385**	**311 385**	**0**	**0**
其他海洋专业技术服务 Other Marine Professional and Technological Services	228 035	228 035	0	0
海洋工程管理服务 Marine Engineering Management Service	83 350	83 350	0	0

6-5 分行业海洋科研机构科技课题情况

Marine Scientific and Technological Research Projects of the Research Institutions by Industry

单位：项 (item)

行　业 Industry	课题数 Number of Research Projects	基础研究 Basic Research	应用研究 Applied Research	试验发展 Experimental Development	成果应用 Result Application	科技服务 Scientific and Technological Service
合　计 Total	**16 331**	**4 276**	**3 735**	**3 997**	**1 516**	**2 807**
海洋基础科学研究 Marine Basic Scientific Research	**12 104**	**4 202**	**3 247**	**2 259**	**853**	**1 543**
海洋自然科学 Marine Natural Science	9 730	3 932	2 932	1 681	439	746
海洋社会科学 Marine Social Science	667	10	20	71	103	463
海洋农业科学 Marine Agricultural Science	1 685	258	291	503	299	334
海洋生物医药 Marine Biomedicine	22	2	4	4	12	0
海洋工程技术研究 Marine Engineering Technology Research	**3 978**	**74**	**469**	**1 682**	**646**	**1 107**
海洋化学工程技术 Marine Chemical Engineeing Technology	1 058	24	157	620	204	53
海洋生物工程技术 Marine Bioengineering Technology	15	0	0	0	11	4

6-5 续表 continued

行 业 Industry	课题数 Number of Research Projects	基础研究 Basic Research	应用研究 Applied Research	试验发展 Experimental Development	成果应用 Result Application	科技服务 Scientific and Technological Service
海洋交通运输工程技术 Marine Communications and Transport Technology	804	7	41	309	199	248
海洋能源开发技术 Marine Energies Development Technology	622	12	80	323	41	166
海洋环境工程技术 Marine Environmental Engineering Technology	241	1	31	70	33	106
河口水利工程技术 Eustuarine Water Conservancy Engineering	1 119	30	154	321	117	497
其他海洋工程技术 Other Marine Engineering Technology	119		6	39	41	33
海洋信息服务业 Marine Information Service	**153**	**0**	**18**	**28**	**2**	**105**
其他海洋信息服务 Other Marine Information Services	153	0	18	28	2	105
海洋技术服务业 Marine Technological Service Industry	**96**	**0**	**1**	**28**	**15**	**52**
其他海洋专业技术服务 Other Marine Professional and Technological Services	16	0	1	11	4	0
海洋工程管理服务 Marine Engineering Management Service	80	0	0	17	11	52

6-6　分行业海洋科研机构科技论著情况
Marine Scientific and Technological Works of the Research Institutions by Industry

行　业 Industry	发表科技论文（篇） Scientific Theses Published (piece)	#国外发表 Published Abroad	出版科技著作（种） Scientific and Technological Works Published (kind)
合计 Total	**16 284**	**5 536**	**384**
海洋基础科学研究 Marine Basic Scientific Research	**11 128**	**4 621**	**189**
海洋自然科学 Marine Natural Science	8 692	4 261	102
海洋社会科学 Marine Social Science	735	3	52
海洋农业科学 Marine Agricultural Science	1 692	357	35
海洋生物医药 Marine Biomedicine	9	0	0
海洋工程技术研究 Marine Engineering Technology Research	**4 732**	**899**	**183**
海洋化学工程技术 Marine Chemical Engineering Technology	884	129	5
海洋生物工程技术 Marine Bioengineering Technology	73	7	4

6-6 续表 continued

行 业 Industry	发表科技论文（篇） Scientific Theses Published (piece)	#国外发表 Published Abroad	出版科技著作（种） Scientific and Technological Works Published (kind)
海洋交通运输工程技术 Marine Communications and Transport Technology	750	136	17
海洋能源开发技术 Marine Energies Development Technology	957	213	25
海洋环境工程技术 Marine Environmental Engineering Technology	164	20	12
河口水利工程技术 Eustuarine Water Conservancy Engineering	1 451	303	86
其他海洋工程技术 Other Marine Engineering Technology	453	91	34
海洋信息服务业 **Marine Information Service**	**333**	**16**	**10**
其他海洋信息服务 Other Marine Information Services	333	16	10
海洋技术服务业 **Marine Technological Service Industry**	**91**	**0**	**2**
其他海洋专业技术服务 Other Marine Professional and Technological Services	12	0	2
海洋工程管理服务 Marine Engineering Management Service	79	0	0

6-7 分行业科研机构科技专利情况
Marine Scientific and Technological Patents of the Research Institutions by Industry

单位：件 (unit)

行 业 Industry	专利申请受理数 Number of Patent Applications Accepted	#发明专利 Patents for Discoveries	专利授权数 Number of Patents Granted	#发明专利 Patents for Discoveries	拥有发明专利总数 Total Number of Patents for Discoveries
合 计 Total	**5 340**	**4 398**	**3 430**	**2 246**	**11 564**
海洋基础科学研究 Marine Basic Scientific Research	**1 980**	**1 478**	**1 304**	**757**	**3 204**
海洋自然科学 Marine Natural Science	1 448	1 156	904	582	2 583
海洋社会科学 Marine Social Science	8	8	2	2	13
海洋农业科学 Marine Agricultural Science	523	313	397	172	607
海洋生物医药 Marine Biomedicine	1	1	1	1	1
海洋工程技术研究 Marine Engineering Technology Research	**3 333**	**2 903**	**2 116**	**1 483**	**8 315**
海洋化学工程技术 Marine Chemical Engineeing Technology	2 556	2 443	1 510	1 275	7 148
海洋生物工程技术 Marine Bioengineering Technology	0	0	0	0	0

6-7 续表 continued

行 业 Industry	专利申请受理数 Number of Patent Applications Accepted	#发明专利 Patents for Discoveries	专利授权数 Number of Patents Granted	#发明专利 Patents for Discoveries	拥有发明专利总数 Total Number of Patents for Discoveries
海洋交通运输工程技术 Marine Communications and Transport Technology	140	56	79	21	156
海洋能源开发技术 Marine Energies Development Technology	359	229	264	91	701
海洋环境工程技术 Marine Environmental Engineering Technology	28	12	29	6	22
河口水利工程技术 Eustuarine Water Conservancy Engineering Technology	90	45	118	44	176
其他海洋工程技术 Other Marine Engineering Technology	160	118	116	46	112
海洋信息服务业 Marine Information Service	**7**	**7**	**0**	**0**	**1**
其他海洋信息服务 Other Marine Information Services	7	7	0	0	1
海洋技术服务业 Marine Technological Service Industry	**20**	**10**	**10**	**6**	**44**
其他海洋专业技术服务 Other Marine Professional and Technological Services	17	9	9	5	41
海洋工程管理服务 Marine Engineering Management Service	3	1	1	1	3

6-8 分行业海洋科研机构R & D情况
R & D in the Marine Scientific Research Institutions by Profession

行业 Industry	R & D人员 (人) R & D Personnel (persons)	R & D经费内部支出 (千元) R & D Internal Expenditure (1 000 yuan)	R & D课题数 (项) Number of R & D Projects (item)
合计 Total	**27 424**	**14 298 108**	**12 008**
海洋基础科学研究 Marine Basic Scientific Research	**18 217**	**8 738 942**	**9 708**
海洋自然科学 Marine Natural Science	16 004	7 969 210	8 545
海洋社会科学 Marine Social Science	280	55 449	101
海洋农业科学 Marine Agricultural Science	1 912	710 785	1 052
海洋生物医药 Marine Biomedicine	21	3 498	10
海洋工程技术研究 Marine Engineering Technology Research	**8 530**	**5 319 259**	**2 225**
海洋化学工程技术 Marine Chemical Engineering Technology	3 435	2 596 392	801
海洋生物工程技术 Marine Bioengineering Technology	0	0	0

6-8 续表 continued

行 业 Industry	R & D人员（人） R & D Personnel (persons)	R & D经费内部支出（千元） R & D Internal Expenditure (1 000 yuan)	R & D课题数（项） Number of R & D Projects (item)
海洋交通运输工程技术 Marine Communications and Transport Technology	1 459	358 550	357
海洋能源开发技术 Marine Energies Development Technology	2 003	1 599 862	415
海洋环境工程技术 Marine Environmental Engineering Technology	352	109 260	102
河口水利工程技术 Eustuarine Water Conservancy Engineering Technology	1 006	503 541	505
其他海洋工程技术 Other Marine Engineering Technology	275	151 654	45
海洋信息服务业 **Marine Information Service**	**462**	**197 535**	**46**
其他海洋信息服务 Other Marine Information Services	462	197 535	46
海洋技术服务业 **Marine Technological Service Industry**	**215**	**42 372**	**29**
其他海洋专业技术服务 Other Marine Professional and Technological Services	81	19 020	12
海洋工程管理服务 Marine Engineering Management Service	134	23 352	17

6-9 分地区海洋科研机构及人员情况
Marine Scientific Research Institutions and Personnel by Regions

地　区 Region	机构数（个） Number of Institutions (unit)	从业人员（人） Employed Population (person)
合 计 Total	**175**	**38 754**
北 京 Beijing	24	13 976
天 津 Tianjin	14	2 646
河 北 Hebei	5	555
辽 宁 Liaoning	17	2 107
上 海 Shanghai	14	4 039
江 苏 Jiangsu	10	2 959
浙 江 Zhejiang	18	1 800
福 建 Fujian	12	1 276
山 东 Shandong	21	3 864
广 东 Guangdong	24	3 250
广 西 Guangxi	9	460
海 南 Hainan	3	215
其 他 Other	4	1 607

6-10 分地区海洋科研机构科技活动人员学历构成
Educational Background Composition of the Personnel Engaged in Scientific and Technological Activities in Marine Scientific Research Institutions by Regions

单位：人 (person)

地区 Region	科技活动人员 Personnel Engaged in Scientifical Activities	博士 Doctor	硕士 Master	大学生 University Graduate	大专生 College Graduate
合计 Total	**32 349**	**7 499**	**9 352**	**10 067**	**3 045**
北京 Beijing	12 371	3 696	3 624	3 055	974
天津 Tianjin	2 192	176	688	968	182
河北 Hebei	525	43	128	260	70
辽宁 Liaoning	1 706	152	465	688	225
上海 Shanghai	3 366	559	1 052	1 189	354
江苏 Jiangsu	1 728	349	485	616	163
浙江 Zhejiang	1 500	149	525	662	120
福建 Fujian	1 224	167	324	406	177
山东 Shandong	3 181	809	884	966	370
广东 Guangdong	2 796	802	756	756	271
广西 Guangxi	371	15	66	213	60
海南 Hainan	175	7	40	61	26
其他 Other	1 214	575	315	227	53

6-11 分地区海洋科研机构科技活动人员职称构成
Professional Title Composition of Personel Engaged in Marine Scientific Research Institutions by Regions

单位：人 (person)

地 区 Region	科技活动人员 Personnel Engaged in Scientifical Activities	高级职称 Senior	中级职称 Intermediate	初级职称 Primary
合 计 Total	**32 349**	**12 380**	**10 928**	**4 998**
北 京 Beijing	12 371	5 274	4 053	1 454
天 津 Tianjin	2 192	779	781	454
河 北 Hebei	525	217	117	29
辽 宁 Liaoning	1 706	604	546	186
上 海 Shanghai	3 366	985	1 041	788
江 苏 Jiangsu	1 728	772	503	286
浙 江 Zhejiang	1 500	541	538	254
福 建 Fujian	1 224	363	433	215
山 东 Shandong	3 181	1 146	1 269	606
广 东 Guangdong	2 796	1 077	933	457
广 西 Guangxi	371	81	160	105
海 南 Hainan	175	23	35	77
其 他 Other	1 214	518	519	87

6-12 分地区海洋科研机构经费收入
Research Funds Receipts of Marine Scientific Research Institutions by Regions

单位：万元 (10 000 yuan)

地 区 Region	经费收入总额 Fund Total	经常费 Routine Fund	科技活动借贷款 Loans in the Scientific and Technological Activities	基本建设中政府投资 Government Investment in the Capital Construction
合 计 Total	**26 556 354**	**24 881 708**	**108 900**	**1 565 746**
北 京 Beijing	10 200 232	9 848 772	0	351 460
天 津 Tianjin	1 550 800	1 402 835	0	147 965
河 北 Hebei	138 739	135 498	0	3 241
辽 宁 Liaoning	1 134 799	1 115 999	0	18 800
上 海 Shanghai	3 072 266	2 980 036	0	92 230
江 苏 Jiangsu	2 085 245	1 936 803	100 000	48 442
浙 江 Zhejiang	1 333 885	1 271 557	5 900	56 428
福 建 Fujian	699 473	599 758	0	99 715
山 东 Shandong	3 247 585	2 648 434	0	599 151
广 东 Guangdong	1 960 673	1 822 651	3 000	135 022
广 西 Guangxi	123 638	123 638	0	0
海 南 Hainan	62 439	59 529	0	2 910
其 他 Other	946 580	936 198	0	10 382

6-13　分地区海洋科研机构科技课题情况

Marine Scientific and Technological Research Projects of the Research Institutions by Regions

单位：项　　(item)

地　区 Region	课题数 Number of Research Projects	基础研究 Basic Research	应用研究 Applied Research	试验发展 Experimental Development	成果应用 Result Application	科技服务 Scientific and Technological Service
合　计 Total	**16 331**	**4 276**	**3 735**	**3 997**	**1 516**	**2 807**
北　京 Beijing	6 045	1 652	1 377	1 212	378	1 426
天　津 Tianjin	723	16	57	335	114	201
河　北 Hebei	94	11	25	35	15	8
辽　宁 Liaoning	339	21	48	162	68	40
上　海 Shanghai	1 127	35	301	411	116	264
江　苏 Jiangsu	1 889	53	411	880	373	172
浙　江 Zhejiang	588	74	93	109	105	207
福　建 Fujian	632	202	230	40	53	107
山　东 Shandong	1 681	559	511	327	185	99
广　东 Guangdong	1 864	661	555	343	42	263
广　西 Guangxi	86	5	12	58	10	1
海　南 Hainan	47	0	0	0	44	3
其　他 Other	1 216	987	115	85	13	16

6-14 分地区海洋科研机构科技论著情况
Marine Scientific and Technological Works of the Research Institutions by Regions

地区 Region	发表科技论文（篇） Scientific Theses Published (piece)	#国外发表 Published Abroad	出版科技著作（种） Scientific and Technological Works Published (kind)
合计 Total	**16 284**	**5 536**	**384**
北京 Beijing	6 238	2 609	144
天津 Tianjin	888	137	30
河北 Hebei	426	6	43
辽宁 Liaoning	418	50	14
上海 Shanghai	1 105	184	39
江苏 Jiangsu	969	307	15
浙江 Zhejiang	588	135	16
福建 Fujian	331	83	3
山东 Shandong	2 094	770	39
广东 Guangdong	1 889	714	32
广西 Guangxi	89	9	2
海南 Hainan	63	1	0
其他 Other	1 186	531	7

6-15 分地区海洋科研机构科技专利情况
Marine Scientific and Technological Patents of the Research Institutions by Regions

单位：件 (unit)

地区 Region	专利申请受理数 Number of Patent Applications Accepted	#发明专利 Patents for Discoveries	专利授权数 Number of Patents Granted	#发明专利 Patents for Discoveries	拥有发明专利总数 Total Number of Patents for Discoveries
合计 Total	**5 340**	**4 398**	**3 430**	**2 246**	**11 564**
北京 Beijing	2 228	1 988	1 435	1 007	5 949
天津 Tianjin	113	68	88	32	153
河北 Hebei	3	3	7		8
辽宁 Liaoning	575	526	324	263	1 544
上海 Shanghai	1 073	853	625	432	1 882
江苏 Jiangsu	172	108	99	47	201
浙江 Zhejiang	115	52	79	28	131
福建 Fujian	53	51	13	11	224
山东 Shandong	455	363	370	226	678
广东 Guangdong	327	238	226	149	527
广西 Guangxi	15	15	14	11	36
海南 Hainan			2	2	2
其他 Other	211	133	148	38	229

6-16 分地区海洋科研机构R & D情况

R & D in the Marine Scientific Research Institutions by Regions

行 业 Industry	R & D人员 （人） R & D Personnel (persons)	R & D经费内部支出 （千元） R & D Internal Expenditure (1 000 yuan)	R & D课题数 （项） Number of R & D Projects (item)
合 计 Total	**27 424**	**14 298 108**	**12 008**
北 京 Beijing	10 283	6 156 253	4 241
天 津 Tianjin	1 549	696 242	408
河 北 Hebei	249	42 408	71
辽 宁 Liaoning	939	634 534	231
上 海 Shanghai	2 727	1 763 624	747
江 苏 Jiangsu	1 695	649 443	1 344
浙 江 Zhejiang	642	331 195	276
福 建 Fujian	790	242 596	472
山 东 Shandong	2 954	1 917 428	1 397
广 东 Guangdong	3 281	1 118 513	1 559
广 西 Guangxi	177	38 588	75
海 南 Hainan			
其 他 Other	2 138	707 284	1 187

主要统计指标解释

1. 海洋科研机构 指有明确的研究方向和任务，有一定水平的学术带头人和一定数量、质量的研究人员，有开展研究工作的基本条件，长期有组织地从事海洋研究与开发活动的机构。

2. 从业人员 指由本机构年末直接组织安排工作并支付工资的各类人员总数。包括固定职工、国家有编制的合同制职工、招聘人员和返聘的离退休人员。不包括离退休人员、停薪留职人员。

3. 从事科技活动人员 指从业人员中的科技管理人员、课题活动人员和科技服务人员。

4. 高级职称 指研究员、副研究员；教授、副教授；高级工程师；高级农艺师；正、副主任医（药、护、技)师；高级实验师；高级统计师；高级经济师；高级会计师；编审(正、副编审)；译审(正、副译审)、高级(主任)记者；正、副研究馆员等。

5. 中级职称 指助理研究员；讲师；工程师；农艺师；主治医(药、护、技)师；实验师；统计师；经济师；会计师；编辑；翻译；记者；馆员等。

6. 初级职称 指研究实习员；助教；助理工程师、技术员；助理农艺师、农业技术员；医(药、护、技)师、医(药、护、技)士；助理实验师、实验员；助理统计师、统计员；助理经济师；助理会计师、会计员；助理编辑、见习编辑；助理翻译；助理记者；助理馆员、管理员等。

7. 科技经费筹集额 指从各种渠道筹集到的计划用于本单位科技活动的经费，不论来源渠道如何。

8. 政府资金 指由各级政府部门直接拨款或企事业单位利用政府资金委托本机构从事科学技术活动所获得的收入。

9. 生产经营活动收入 指本机构在科研、技术等专业业务活动以外开展非独立核算的经营活动取得的收入，包括产品（商品）销售收入、经营服务收入、工程承包收入、租赁收入和其他经营收入。

10. 其他收入 指开展科技活动与生产经营活动以外的各项活动的收入，包括：用于离退休人员的政府拨款。

11. 非科技活动借贷款 指本机构为开展非科技活动从各种渠道获得的各类借、贷款。不论偿还形式、期限和数额如何，均按当年获得的借、贷款额填报。不包括基本建设贷款。

12. 基础研究 为获得新知识而进行的独创性研究。其目的是揭示观察到的现象和事实的基本原理和规律，而不以任何特定的实际应用为目的。

13. 应用研究 为获得新的科学技术知识而进行的独创性研究。它主要针对某一特定的实际应用目的。应用研究通常是为了确定基础研究成果或知识的可能的用途，或是为达到某一具体的、预定的实际目的确定新的方法(原理性)或途径。

14. 试验发展 利用从研究或实际经验获得的知识，为生产新的材料、产品和装置，建立新的工艺和系统，以及对已生产或建立的上述各项进行实质性的改进而进行的系统性工作。

15. 成果应用 为解决 R&D 活动阶段产生的新产品、新装置、新工艺、新技术、新方法、新系统和服务等能投入生产或在实际应用中所存在的技术问题而进行的系统性活动。它不具有创新成分。此类活动包括为达到生产目的而进行的定型设计和试制以及为扩大新产品的生产规模和探索新方法、新技术、新工艺等的应用领域而进行的适应性试验。

16. 科技服务 与科学研究与实验发展有关，并有助于科学技术知识的产生、传播和应用的活动。包括为扩大科技成果的使用范围而进行的示范性推广工作；为用户提供科技信息和文献服务的系统性工作；为用户提供可行性报告、技术方案、建议及进行技术论证等技术咨询工作；自然、生物现象的日常观测、监测，资源的考查和勘探；有关社会、人文、经济现象的通用资料的收集，如统计、市场调查等以及这些资料的常规分析与整理；为社会和公众提供的测试、标准化、计量、计算、质量控制和专利服务，不包括工商企业为进行正常生产而开展的上述活动。

17. 生产性活动 由于业务特殊的工艺设备条件，或掌握某种技术专长或诀窍，所进行的小量非常规生产。

18. 科技论文 在全国性学报或学术刊物上、省部属大专院校对外正式发行的学报或学术刊物上发表的论文以及向国外发表的论文。

19. 科技著作 经过正式出版部门编印出版的科技专著、大专院校教科书、科普著作。

20. 专利申请受理数 当年本单位向专利管理部门提出申请并被受理的职务专利申请件数。

21. 专利授权数 当年由专利管理部门授予本单位专利权的职务专利件数。

Explanatory Notes on Main Statistical Indicators

1. Marine Scientific Research Institution refers to the institution which has definite research orientations and tasks, high-level academic leading personnel and fair-sized, qualified research personnel, and basic conditions for research work and which is engaged for a long time in the marine research and development activities in an organized way.

2. Employees refer to the total number of personnel of various kinds employed and paid by the institution at the end of the year, including fixed employees, contract workers of staff belonging to the state authorized staff , recruited personnel, reemployed retired personnel, but not including the retired and the personnel on leave with pay suspension.

3. Personnel Engaged in Scientific and Technological Activities refers to the personnel for scientific and Technological management, personnel engaged in the activities of research topics and scientific and technological service personnel.

4. Senior Technical Title refer to research scientist, associate research scientist; professor, associate professor; senior engineer; senior agronomist; professor-rank and associate professor-rank doctor (pharmacists, nurses and technicians); senior laboratory technician; senior statisticians; senior economic engineer; chief accountant; senior editor (professor and associate professor ranks); senior translator (professor and associate professor ranks); senior journalist; research librarian (professor and associate professor ranks), etc.

5. Intermediate Technical Title refer to assistant research scientist; lecturer; engineer; agronomist; lecturer-rank doctor (pharmacist, nurse and technician); laboratory technician; lecturer-rank statistician;

economic engineer; accountant; editor; translator; journalist; librarian, etc.

6. Primary Technical Title refers to trainee researcher; assistant; assistant engineer, technician, assistant agronomist, agricultural technician; assistant-rank doctor (pharmacist, nurse and technician); assistant laboratory technician; assistant statistician; assistant economic engineer, assistant accountant; assistant editor, editor on probation; assistant translator; assistant journalist; assistant research librarian, librarian ,etc.

7. Amount of Scientific and Technological Funds Raised refers to the funds raised through all channels planned to be used as funds for the scientific and technological activities in the institution regardless of their source and channels.

8. Funds from government refers to the direct appropriations by the government departments at all levels or the earnings from conducting scientific and technological activities entrusted to the institution by enterprises as institutions by earning the funds from government.

9. Earnings from Production as Business Activities refer to the incomes obtained from the non-independent accounting business activities carried out by the institution beyond the scientific research, technological and professional activities, including these from sale of products(goods), business and service, contracted projects, leasing and other business.

10. Other Incomes refer to those from the various activities conducted other than scientific and technological activities and production and business activities, including the government appropriations for retired personnel.

11. Loan for Non-scientific as Technological Activities refers to the various types loan obtained by the institutions through all channels for carrying out non-scientific and technological activities, not including the load for capital construction. The loan, irrespective of its form of reimbursement, term and amount is filled in a form and submitted to the authorities as the amount acquired in the current year.

12. Basic Research refers to the original research to acquire new knowledge. It is aimed at revealing the basic principles and laws of the phenomena and facts observed, but not at any specific practical applications.

13. Applied Research refers to the original research to acquire new scientific and technological knowledge. It mainly serves the purpose of a particular practical application. The purpose of applied research is usually to define the potential uses of the research finds or knowledge obtained from basic research or to identify new methods (principles) or ways to reach a specific and predetermined goal.

14. Experimental Development refers to the systematic work carried out to establish new technologies and systems for producing new materials, products and equipment by using the knowledge obtained from research or practical experience, or to make substantial improvement of the above-mentioned which have been produced or established.

15. Result Application refers to the systematic activities carried out to solve the technical problems that might crop up in the production or practical application of the new products, devices, technologies, techniques, methods, systems and service occurring in the course of R & D activities.

They do not bring forth new ideas. Such activities include the finalizing design and trial-production for the purpose of production as well as the adaptive tests to expand the production scale of new products and the application areas of new methods, techniques and technologies.

16. Scientific and Technological Service refers to the activities that are associated with the scientific research and experimental development, and contribute to the generation, dissemination and application of scientific and technological knowledge, which include the demonstrative work of popularization to enlarge the use scope of scientific and technological achievements; the systematic work of providing the users with scientific and technological information and literature service; the technical consultation work of providing users with feasibility reports, technical schemes and recommendations and carrying out technical demonstration; routine observation and monitoring of natural and biological phenomena, and the survey and exploration of resources; collection of universal data on the appropriate social, cultural and economic phenomena, such as statistics and market survey, as well as the routine analysis and sorting-out of these data; the provision for the society and the public of such service as testing, standardization, computation, quality control and patent, but not including the type of the above-mentioned activities carried out by industrial and commercial enterprises for the purpose of normal production.

17. Productive Activity refers to the small-scale and non-conventional productions due to the presence of special technologies and equipment or mastery of a particular technical expertise or secret of success.

18. Scientific Treatises refer to the theses published in the national journals or academic publications, those officially issued journals or academic publications by universities and colleges under provinces or ministries as well as theses published abroad.

19. Scientific and Technological Works refer to the scientific and technological monographs, textbooks for universities and colleges and popular science books published by the official publishing houses.

20. Number of Patents Applied and Accepted refers to the number of professional patent applications of the unit to the patent administrative department and accepted by it in the year.

21. Number of Patents Granted refers to the number of the professional patents granted to the unit by the patent administrative department in the year.

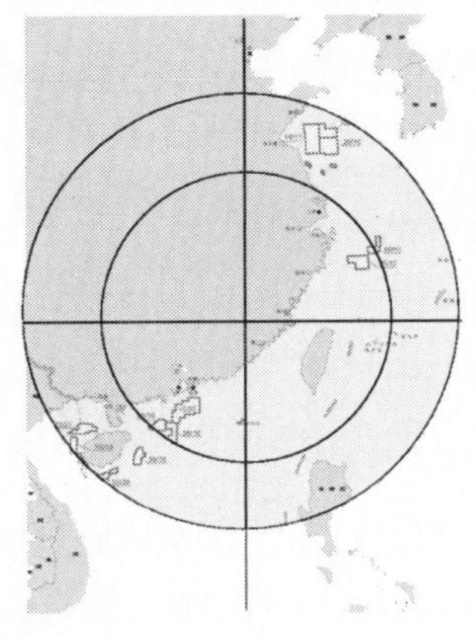

7

海 洋 教 育

Marine Education

7-1 全国各海洋专业博士研究生情况
Doctoral Students from Marine Specialities

专业 Speciality	专业点数（个） Number of Speciality Agencies (unit)	学生数（人） Number of Students (person)			
		毕业生 Graduates	招生 Entrants	在校生 Enrollment	预计毕业生数 Estimated Graduates of Next Year
合计 Total	**137**	**673**	**929**	**4 031**	**1 937**
物理海洋学 Physical Oceanography	5	37	66	258	128
海洋化学 Marine Chemistry	5	27	34	137	75
海洋生物学 Marine Biology	8	69	77	263	112
海洋地质 Marine Geology	9	29	39	234	140
海洋科学新专业 New Speciality of Marine Science	6	66	119	403	191
港口海岸及近海工程 Coastal Harbour and Offshore Engineering	9	51	63	297	120
船舶与海洋结构物设计制造 Ships and Marine Structures Design and Manufacture	11	53	80	358	139
轮机工程 Turbine Engineering	8	31	48	276	94

7-1 续表 continued

专 业 Speciality	专业点数（个） Number of Speciality Agencies (unit)	学生数（人） Number of Students (person)			
		毕业生 Graduates	招 生 Entrants	在校生 Enrollment	预计毕业生数 Estimated Graduates of Next Year
水声工程 Hydroacoustic Engineering	7	22	32	163	95
船舶与海洋工程新专业 New Speciality of Shipand Marine Engineering	5	29	67	226	116
捕捞学 Science of Fishing	2	5	4	14	7
航空、航天与航海医学 Aeronautical, Aerospace and Nautical Medicine	1	0	1	1	0
水产品加工及贮藏工程 Aquatic Product Sprocessing and Storing Engineering	9	10	1	36	27
水产新专业 New Specialities of Aquaculture	5	15	29	78	38
水产养殖 Aquaculture	6	45	52	230	100
水生生物学 Hydrobiology	19	72	87	349	161
水文学及水资源 Hydrology and Water Resource	18	101	110	636	357
渔业资源 Fishery Resource	4	11	20	72	37

7-2 全国各海洋专业硕士研究生情况
Postgraduate Students from Marine Specialities

专 业 Speciality	专业点数（个） Number of Speciality Agencies (unit)	学生数（人） Number of Students (person)			
		毕业生 Graduates	招 生 Entrants	在校生 Enrollment	预计毕业生数 Estimated Graduates of Next Year
合 计 Total	**345**	**3 356**	**3 491**	**10 170**	**3 392**
物理海洋学 Physical Oceanography	16	78	182	408	93
海洋化学 Marine Chemistry	17	89	120	329	94
海洋生物学 Marine Biology	24	342	273	924	364
海洋地质 Marine Geology	19	126	158	449	147
海洋科学新专业 New Speciality of Marine Science	9	83	244	560	112
港口海岸、及近海工程 Coastal Harbour and Offshore Engineering	23	312	280	845	311
船舶与海洋结构物设计制造 Ship and Marine Structures Design and Manufacture	22	349	309	1 059	380
轮机工程 Turbine Engineering	14	290	233	658	209

7-2 续表 continued

专 业 Speciality	专业点数（个） Number of Speciality Agencies (unit)	学生数（人） Number of Students (person)			
		毕业生 Graduates	招 生 Entrants	在校生 Enrollment	预计毕业生数 Estimated Graduates of Next Year
水声工程 Hydroacoustic Engineering	9	143	122	358	113
船舶与海洋工程新专业 New Specialities in Shipping and Marine Engineering	15	90	226	477	109
捕捞学 Science of Fishing	5	25	27	86	33
航空、航天与航海医学 Aeronautical, Aerospace and Nautical Medicine	2	15	0	30	16
水产品加工及贮藏工程 Aquatic Products Processing and Storing Engineering	25	79	97	303	103
水产新专业 New Specialities of Aquaculture	5	30	82	120	23
水产养殖 Aquaculture	31	401	417	1 288	448
水生生物学 Hydrobiology	44	355	235	830	333
水文学及水资源 Hydrology and Water Resource	55	471	402	1 186	412
渔业资源 Fishery Resource	10	78	84	260	92

7-3 全国普通高等教育各海洋专业本科学生情况
Undergraduates from Marine Specialities in the National Ordinary Higher Education

专业 Speciality	专业点数（个） Number of Speciality Agencies (unit)	学生数（人） Number of Students (person)			
		毕业生 Graduates	招生 Entrants	在校生 Enrollment	预计毕业生数 Estimated Graduates of Next Year
合计 Total	**241**	**14 730**	**17 555**	**65 668**	**15 384**
海洋科学 Marine Science	25	1 068	973	4 264	1 118
海洋技术 Marine Technology	20	732	561	2 595	638
海洋资源与环境 Marine Living Resources and Environment	12	174	364	829	93
海洋科学类专业 New Specialties Under the Category of Marmine Sciences	8	62	493	905	53
港口航道与海岸工程 Harbour Channel and Coastal Engineering	25	1 697	1 476	7 124	1 942
船舶与海洋工程 Water Resource and Ocean Engineering	29	2 882	2 271	11 046	3 066
航海技术 Nautical Technology	17	2 218	2 673	9 823	2 254
轮机工程 Turbine Engineering	21	2 839	3 180	12 073	2 838
海事管理 Maritime Affairs Manegernent	2	109	134	488	125
船舶与海洋工程 Ship and Marine Engineering	29	2 882	2 271	11 046	3 066
水产养殖学 Aquaculture	51	2 380	3 591	11 884	2 669
海洋渔业科学与技术 Science and Technology of Marine Fishery	9	373	504	1 762	369
水族科学与技术 Science and Technology of Aquatic Animals	7	196	318	1 038	198
海洋工程与技术 Ocean Engineering and Technology	5	0	439	985	21
海洋资源开发技术 Marine Resources Exploitation Technology	8	0	255	529	0
海洋工程类专业 New Speciality of Marine Engineering	2	0	323	323	0

7-4 全国普通高等教育各海洋专业专科学生情况
Students from Marine Specialities of the Colleges for Professional Training in the National Ordinary Higher Education

专业 Speciality	专业点数（个） Number of Speciality Agencies (unit)	学生数（人） Number of Students (person)			
		毕业生 Graduates	招生 Entrants	在校生 Enrollment	预计毕业生数 Estimated Graduates of Next Year
合计 Total	**505**	**36 150**	**28 352**	**93 990**	**34 321**
港口工程技术 Harbour Engineering	11	575	506	1 647	538
港口业务管理 Harbour Business Management	19	1 084	846	2 912	1 165
港口与航运管理 Harbour and Shipping Management	12	333	570	1 487	438
港口机械应用技术 Applied Harbour Machinery Technology	3	235	197	697	215
港口物流设备与自动控制 Harbour Logistics Equipment and Automatic Control	20	1 049	911	2 997	1 035
集装箱运输管理 Container Transportation Management	20	1 042	725	2 465	999
国际航运业务管理 International Shipping Business Management	34	1 876	2 253	6 541	2 004
港口运输类专业 Port Transportation Type of Majors	3	0	211	212	0
海关管理 Customs Management	3	12	66	133	14
报关与国际货运 Customs Clearing and International Freight Transport	211	15 591	11 351	37 504	13 470
轮机工程技术 Engine Engineering Technology	47	6 299	4 658	15 728	6 009
船舶检验 Ship Inspection	7	99	111	360	181
船舶舾装 Ship Equipment and Installations	5	115	100	388	184

7-4 续表 continued

专 业 Speciality	专业点数（个） Number of Speciality Agencies (unit)	学生数（人） Number of Students (person)			
		毕业生 Graduates	招 生 Entrants	在校生 Enrollment	预计毕业生数 Estimated Graduates of Next Year
船艇动力管理 Ship and Boat Power Equipment Management	1	68	0	48	47
船舶工程技术 Ship Engineering Technology	39	5 567	3 667	14 155	5 727
船机制造与维修 Ship Engines Manufacturing and Maintenance	7	291	372	1 253	482
水运管理 Water Transport Management	3	120	93	350	141
水信息技术 Hydrological Information Technology	2	55	0	51	43
水环境监测与保护 Water Environmental Monitoring and Protection	8	279	300	639	192
水政水资源管理 Water Resources Management by Water Administration	4	47	222	422	95
水产养殖技术 Aquaculture Technology	33	1 125	1 030	3 543	1 160
水产养殖类专业 Aquaculture Type of Majors	1	0	18	18	0
渔业综合技术 Integrated Fishery atechnology	4	64	8	20	12
水生动植物保护 Aquatic Animals and Plants Conservation	2	64	0	16	16
水上运输类专业 Water Transportation Type of Majors	1	0	0	1	1
港口航道与治河工程 Harbour, Channel and River Harnessing Engineering	4	160	109	357	153
水文自动化测报技术 New Specialties Under the Category of Marmine Sciences	1	0	28	46	0

7-5 全国成人高等教育各海洋专业本科学生情况 Undergraduates from Marine Specialities in the National Adult Higher Education

专 业 Speciality	专业点数（个） Number of Speciality Agencies (unit)	学生数（人） Number of Students (person)			
		毕业生 Graduates	招 生 Entrants	在校生 Enrollment	预计毕业生数 Estimated Graduates of Next Year
合 计 Total	**54**	**2 138**	**2 631**	**6 880**	**2 864**
海洋科学 Marine Science	1	2	0	0	0
海洋技术 Marine Technology	1	96	46	62	0
港口航道与海岸工程 Harbour Channel and Coastal Engineering	2	24	17	102	60
航海技术 Navigation Technology	9	208	199	632	266
轮机工程 Turbine Engineering	9	180	348	978	420
海事管理 Maritime Affairs Management	1	11	0	0	0
船舶与海洋工程 Ship and Marine Engineering	14	1 074	1 261	3 558	1 553
水产养殖学 Aquaculture	17	543	760	1 548	565

7-6　全国成人高等教育各海洋专业专科学生情况

Students from Marine Specialities of the Colleges for Professional Training in the National Adult Higher Education

专　业 Speciality	专业点数（个） Number of Speciality Agencies (unit)	学生数（人） Number of Students (person)			
		毕业生 Graduates	招 生 Entrants	在校生 Enrollment	预计毕业生数 Estimated Graduates of Next Year
合　计 Total	**140**	**6 886**	**4 813**	**16 519**	**7 695**
港口业务管理 Harbour Business Management	1	11	0	0	0
港口与航运管理 Harbour and Shipping Management	1	1	0	32	31
国际航运业务管理 International Shipping Business Management	4	40	43	309	72
港口物流设备与自动控制 Harbour Logistics Equipment and Automatic Control	1	149	48	98	0
轮机工程技术 Engine Engineering	32	3 544	1 711	8 717	4 533
船舶检验 Ship Inspection	3	18	4	18	7
船舶工程技术 Ship Engineering	23	1 966	1 895	4 333	1 807
船机制造与维修 Ship Engines Manufacturing and Maintenance	3	1	4	184	72

7-6 续表 continued

专 业 Speciality	专业点数（个） Number of Speciality Agencies (unit)	学生数（人） Number of Students (person)			
		毕业生 Graduates	招 生 Entrants	在校生 Enrollment	预计毕业生数 Estimated Graduates of Next Year
船艇动力管理 Ship and Boat Power Equipment Management	1	29	0	48	48
海关管理 Customs Management	2	1	0	1	1
报关与国际货运 Customs Declaration/International Freight Transport	37	941	790	2 108	881
水政水资源管理 Water Resources Management by Water Administration	5	35	22	62	25
水上运输类新专业 New Specialties under the Category of Water Transport	2	0	38	110	35
渔业综合技术 Fishery Ingrated Technology	1	28	0	2	2
水产养殖技术 Aquaculture Technology	23	120	254	490	181
水产养殖类新专业 New Specialties under the Category of Aquaculture	1	2	4	7	0

7-7 全国中等职业教育各海洋专业学生情况
Students from Marine Specialities in the National Secondary Vocational Education

专业 Speciality	专业点数（个） Number of Speciality Agencies (unit)	学生数（人） Number of Students (person)			
		毕业生 Graduates	招生 Entrants	在校生 Enrollment	预计毕业生数 Estimated Graduates of Next Year
合计 Total	**461**	**30 612**	**16 773**	**58 835**	**26 681**
海水生态养殖 Seawater Ecological Cultivation	23	1 131	605	3 718	1 796
农林牧渔类新专业 New Specialities of Agriculture, Forestry, Animal Husbaudry and Fishery	67	7 325	2 825	11 638	4 776
航海捕捞 Sea Fishing	10	365	30	277	179
水文与水资源勘测 Hydrological and Water Resources Survey	4	112	188	330	128
风电场机电设备运行与维护 Operation and Maintenance of Electromechanical Equipment in	46	2 611	1 083	3 963	1 781
船舶制造与修理 Ships Building and Repair	64	3 042	2 760	8 137	3 495
船舶机械装置安装与维修 Installation and Maintenance of Ships' Mechanical Equipment	14	663	427	1 383	458
船舶驾驶 Ship Piloting	78	7 272	3 865	12 977	6 623

7-7 续表 continued

专 业 Speciality	专业点数（个） Number of Speciality Agencies (unit)	学生数（人） Number of Students (person)			
		毕业生 Graduates	招 生 Entrants	在校生 Enrollment	预计毕业生数 Estimated Graduates of Next Year
轮机管理 Engines Management	61	5 446	2 650	8 859	4 282
船舶水手与机工 Ship Sailors and Mechanics	29	561	787	2 460	1 176
船舶电气技术 Ship Electric Technology	23	551	417	1 223	559
船舶通信与导航 Marine Communication and Navigation	1	0	74	74	74
外轮理货 Foreign Ships Freight Forwarding	14	475	214	751	186
船舶检验 Ships Inspection	4	229	12	372	313
港口机械运行与维护 Operation and Maintenance of Harbour Machinery	21	808	790	2 597	833
工程潜水 Engineering Diving	2	21	46	76	22

注：此表不包含技工学校相关数据。
Note: Data related to Technical Schools are not included in the table.

7-8 分地区各海洋专业博士研究生情况
Doctoral Students in Marine Specialities by Regions

地 区 Region	专业点数（个） Number of Speciality Agencies (unit)	学生数（人） Number of Students (person)			
		毕业生 Graduates	招 生 Entrants	在校生 Enrollment	预计毕业生数 Estimated Graduates of Next Year
合 计 Total	**137**	**673**	**929**	**4 031**	**1 937**
北 京 Beijing	7	39	23	112	56
天 津 Tianjin	5	6	5	25	15
辽 宁 Liaoning	8	48	52	331	44
上 海 Shanghai	16	43	97	408	220
江 苏 Jiangsu	13	60	93	459	230
浙 江 Zhejiang	5	5	28	81	19
福 建 Fujian	6	28	39	141	69
山 东 Shandong	19	220	263	1 118	611
广 东 Guangdong	10	49	70	242	104
广 西 Guangxi	1	0	1	1	0
海 南 Hainan	1	0	2	3	0
其 他 Others	46	175	256	1 110	569

7-9 分地区各海洋专业硕士研究生情况
Postgraduate Students in Marine Specialities by Regions

地 区 Region	专业点数（个） Number of Speciality Agencies (unit)	学生数（人） Number of Students (person)			
		毕业生 Graduates	招 生 Entrants	在校生 Enrollment	预计毕业生数 Estimated Graduates of Next Year
合 计 Total	**345**	**3 356**	**3 491**	**10 170**	**3 392**
北 京 Beijing	21	105	110	310	101
天 津 Tianjin	11	68	86	215	62
河 北 Hebei	6	16	8	27	15
辽 宁 Liaoning	23	404	357	1 038	306
上 海 Shanghai	30	411	423	1 166	404
江 苏 Jiangsu	24	390	371	1 120	380
浙 江 Zhejiang	26	153	216	581	180
福 建 Fujian	22	160	206	603	179
山 东 Shandong	37	398	466	1 419	488
广 东 Guangdong	26	227	244	721	233
广 西 Guangxi	4	20	14	52	22
海 南 Hainan	3	28	25	76	28
其 他 Others	112	976	965	2 842	994

7-10 分地区普通高等教育各海洋专业本科学生情况
Undergraduates in the Marine Specialities of Ordinary Higher Education by Regions

地 区 Region	专业点数（个） Number of Speciality Agencies (unit)	学生数（人） Number of Students (person)			
		毕业生 Graduates	招 生 Entrants	在校生 Enrollment	预计毕业生数 Estimated Graduates of Next Year
合 计 Total	**241**	**14 730**	**17 555**	**65 668**	**15 384**
北 京 Beijing	3	160	98	491	145
天 津 Tianjin	12	756	825	3 462	870
河 北 Hebei	10	230	429	1 081	213
辽 宁 Liaoning	22	2 157	2 163	8 814	2 276
上 海 Shanghai	14	1 169	1 460	5 084	1 110
江 苏 Jiangsu	29	1 412	1 450	5 905	1 444
浙 江 Zhejiang	29	625	1 018	3 631	760
福 建 Fujian	13	1 026	1 560	5 556	1 097
山 东 Shandong	30	1 925	2 229	7 972	1 714
广 东 Guangdong	16	833	1 567	4 403	895
广 西 Guangxi	6	136	281	884	187
海 南 Hainan	2	103	153	549	125
其 他 Others	55	4 198	4 322	17 836	4 548

7-11 分地区普通高等教育各海洋专业专科学生情况
Students from of the Marine Specialities Colleges for Professional Training in the Ordinary Higher Education by Regions

地区 Region	专业点数（个） Number of Speciality Agencies (unit)	学生数（人） Number of Students (person)			
		毕业生 Graduates	招生 Entrants	在校生 Enrollment	预计毕业生数 Estimated Graduates of Next Year
合计 Total	**505**	**36 150**	**28 352**	**93 990**	**34 321**
北京 Beijing	1	0	22	22	0
天津 Tianjin	13	1 578	1 177	3 926	1 441
河北 Hebei	29	1 366	637	2 092	863
辽宁 Liaoning	34	2 443	3 121	9 864	3 015
上海 Shanghai	32	2 565	1 747	5 932	2 234
江苏 Jiangsu	58	5 561	3 156	12 090	4 824
浙江 Zhejiang	32	2 167	1 809	5 842	1 984
福建 Fujian	32	1 928	1 739	5 179	1 778
山东 Shandong	68	6 258	4 361	15 319	5 817
广东 Guangdong	33	1 731	2 303	6 345	2 109
广西 Guangxi	26	1 266	1 148	3 544	1 230
海南 Hainan	9	519	354	1 148	347
其他 Others	167	10 134	7 415	24 779	9 542

7-12 分地区成人高等教育各海洋专业本科学生情况
Students from Marine Specialities in the Adult Higher Education by Regions

地 区 Region	专业点数（个） Number of Speciality Agencies (unit)	学生数（人） Number of Students (person)			
		毕业生 Graduates	招 生 Entrants	在校生 Enrollment	预计毕业生数 Estimated Graduates of Next Year
合 计 Total	**54**	**2 138**	**2 631**	**6 880**	**2 864**
天 津 Tianjin	2	78	67	182	77
河 北 Hebei	1	17	11	40	20
辽 宁 Liaoning	7	267	288	689	401
上 海 Shanghai	4	124	252	362	30
江 苏 Jiangsu	3	787	913	2 495	1 126
浙 江 Zhejiang	6	38	13	307	59
福 建 Fujian	1	8	18	55	21
山 东 Shandong	10	123	48	235	125
广 东 Guangdong	6	27	171	650	274
广 西 Guangxi					
其 他 Others	14	669	850	1 865	731

7-13 分地区成人高等教育各海洋专业专科学生情况
Students from Marine Specialities of the colleges for Professional Training in the Adult Higher Education by Regions

地 区 Region	专业点数（个） Number of Speciality Agencies (unit)	学生数（人） Number of Students (person)			
		毕业生 Graduates	招 生 Entrants	在校生 Enrollment	预计毕业生数 Estimated Graduates of Next Year
合 计 Total	**140**	**6 886**	**4 813**	**16 519**	**7 695**
天 津 Tianjin	3	113	65	391	141
河 北 Hebei	2	22	3	15	12
辽 宁 Liaoning	18	1 863	1 243	4 001	2 026
上 海 Shanghai	12	503	459	1 667	768
江 苏 Jiangsu	18	1 074	1 460	3 304	1 370
浙 江 Zhejiang	13	1 037	229	1 908	1 364
福 建 Fujian	5	262	17	398	227
山 东 Shandong	15	418	111	1 034	553
广 东 Guangdong	12	619	216	691	60
广 西 Guangxi	4	2	3	10	7
海 南 Hainan	1	0	0	1	0
其 他 Others	37	973	1 007	3 099	1 167

7-14 分地区中等职业教育各海洋专业学生情况
Students from Marine Specialities in the Secondary Vocational Education by Regions

地 区 Region	专业点数（个） Number of Speciality Agencies (unit)	学生数（人） Number of Students (person)			
		毕业生 Graduates	招 生 Entrants	在校生 Enrollment	预计毕业生数 Estimated Graduates of Next Year
合 计 Total	**461**	**30 612**	**16 773**	**58 835**	**26 681**
天 津 Tianjin	7	496	155	606	195
河 北 Hebei	26	3 094	801	3 625	1 570
辽 宁 Liaoning	50	1 493	1 589	5 417	2 542
上 海 Shanghai	18	1 487	1 070	3 650	1 371
江 苏 Jiangsu	51	2 722	980	4 713	2 754
浙 江 Zhejiang	23	1 292	577	2 837	1 411
福 建 Fujian	43	3 518	1 212	5 554	2 617
山 东 Shandong	57	5 224	2 913	8 494	4 001
广 东 Guangdong	20	771	462	1 933	1 274
广 西 Guangxi	10	398	959	3 378	1 372
海 南 Hainan	7	171	240	436	248
其 他 Others	149	9 946	5 815	18 192	7 326

7-15 分地区开设海洋专业高等学校教职工数
Number of Teaching and Administrative Staff in the Universities and Colleges Offering Marine Specialities by Regions

地 区 Region	学校（机构）数（个） Number of Colleges (Institutions) (unit)	教职工数（人） Number of Teaching and Administrative Staff (person)	专任教师数（人） Number of Full-Time Teachers (person)
合 计 Total	**393**	**471 827**	**299 057**
北 京 Beijing	4	5 689	3 450
天 津 Tianjin	12	15 865	10 688
河 北 Hebei	22	20 557	12 682
辽 宁 Liaoning	23	20 319	12 630
上 海 Shanghai	16	21 262	10 440
江 苏 Jiangsu	44	67 757	43 686
浙 江 Zhejiang	22	28 984	17 328
福 建 Fujian	20	20 824	13 133
山 东 Shandong	45	52 382	35 016
广 东 Guangdong	23	32 818	20 796
广 西 Guangxi	16	14 108	9 930
海 南 Hainan	6	6 600	3 732
其 他 Others	140	164 662	105 546

主要统计指标解释

海洋专业　指高等教育和中等职业教育所设的与海洋有关的专业。

Explanatory Notes on Main Statistical Indicators

Marine Speciality refers to the ocean-related speciality in higher education and the professional secondary vocational education.

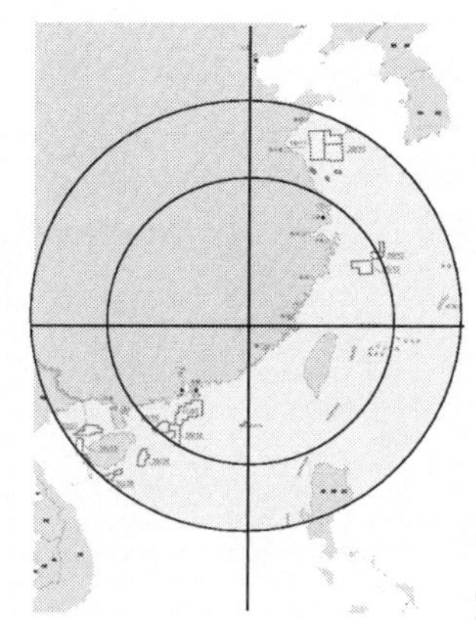

8

海洋环境保护

Marine Environmental Protection

8-1 海区海水水质评价结果
Seawater Quality Assessment Results

项　目 Item	全国总计 National Total	渤　海 Bohai Sea	黄　海 Huanghai Sea	东　海 East China Sea	南　海 South China Sea
第二类水质海域面积（平方千米） Area of the Second Grade Sea Waters (km^2)	**47 160**	9 060	16 010	13 640	8 450
第三类水质海域面积（平方千米） Area of the Third Grade Sea Waters (km^2)	**36 490**	12 920	10 590	8 600	4 380
第四类水质海域面积（平方千米） Area of the Fourth Grade Sea Waters (km^2)	**15 630**	2 930	4 710	5 790	2 200
劣于第四类海域面积（平方千米） Sea Area of the Sea Waters Inferior to the Fourth Grade (km^2)	**44 340**	8 490	3 500	24 820	7 530
首要超标污染物 Prime Pollutants Exceeding the Set Standard	**无机氮、活性磷酸盐、石油类 Inorganic Nitrogen, Active Phosphate, Oils**	无机氮、活性磷酸盐 Inorganic Nitrogen, Active Phosphate	无机氮、活性磷酸盐 Inorganic Nitrogen, Active Phosphate	无机氮、活性磷酸盐 Inorganic Nitrogen, Active Phosphate	无机氮、活性磷酸盐、石油类 Inorganic Nitrogen, Active Phosphate, Oils

8-2 海区废弃物海洋倾倒情况
Ocean Dumping of Wastes by Sea Area

海 区 Sea Area	疏浚物 （万立方米） Dredged Materials (10 000m^3)	惰性无机地质废料 （立方米） Inert, Inorganic Geologic Wastes (m^3)
合 计 **Total**	**13 453.46**	**6 322 619**
渤黄海 Bohai and Huanghai Sea	77.30	3 800
东 海 East China Sea	9 716.00	
南 海 South China Sea	3 660.16	6 318 819

8-3 海区海洋石油勘探开发污染物排放入海情况
Discharge of Pollutants into the Sea from Offshore Oil Exploration and Exploitation

海 区 Sea Area	生产污水 （万立方米） Production Sewage (10 000m^3)	泥浆 （立方米） Sludge (m^3)	钻屑 （立方米） Debris from Drilling (m^3)	机舱污水 （立方米） Sewage from Engineroom (m^3)	食品废弃物 （立方米） Food Wastes (m^3)	生活污水 （立方米） Domestic Sewage (m^3)
合 计 Total	**14 792.88**	**77 708.70**	**69 127.80**	**3 820.90**	**11 151.37**	**445 556.70**
渤 海 Bohai Sea	626.92	9 594.00	38 936.00		9 814.00	148 416.00
黄 海 Huanghai Sea						
东 海 East China Sea	65.20	4 034.30	4 599.40		868.00 *	43 000.00 *
南 海 South China Sea	14 100.76	64 080.40	25 592.40	3 820.90	469.37 *	254 140.70 *

注：*单位为吨。

Notes:* unit is ton.

8-4 沿海地区工业废水排放及处理情况
Discharge and Treatment of Industrial Waste Water by Coastal Regions

单位：万吨 (10 000 t)

地区 Region		工业废水排放总量 Total Volume of Industrial Waste Water Discharged	直排入海 Discharged Directly into the Sea	工业废水处理量 Industrial Waste Water Treated
总计 Total		**1 189 064.60**	**106 885.26**	**2 845 694.86**
环渤海经济区 Round-the-Bohai Sea Economic Zone	**合计 Total**	**388 032.61**	**31 182.15**	**1 355 272.51**
	辽宁 Liaoning	78 285.60	21 102.53	213 534.10
	河北 Hebei	109 875.99	1 157.52	766 321.66
	天津 Tianjin	18 691.93	43.40	29 160.99
	山东 Shandong	181 179.09	8 878.70	346 255.76
长江三角洲经济区 Yangtze River Delta Economic Zone	**合计 Total**	**429 658.70**	**18 231.01**	**717 774.47**
	江苏 Jiangsu	220 558.62	723.14	395 898.52
	上海 Shanghai	45 425.76	9 837.36	65 019.18
	浙江 Zhejiang	163 674.32	7 670.51	256 856.77
海峡西岸经济区 Economic Zone on the West Side of the Taiwan Straits	**合计 Total**	**104 657.99**	**42 496.59**	**168 166.99**
	福建 Fujian	104 657.99	42 496.59	168 166.99
珠江三角洲经济区 Zhujiang River Delta Economic Zone	**合计 Total**	**170 462.59**	**8 730.23**	**293 036.91**
	广东 Guangdong	170 462.59	8 730.23	293 036.91
环北部湾经济区 Round-the-Beibu Gulf Economic Zone	**合计 Total**	**96 252.71**	**6 245.28**	**311 443.98**
	广西 Guangxi	89 508.46	2 995.57	304 910.86
	海南 Hainan	6 744.25	3 249.71	6 533.12

8-5 沿海城市工业废水排放及处理情况
Discharge and Treatment of Industrial Waste Water by Coastal Cities

单位：万吨 (10 000 t)

沿海城市 Coastal City	工业废水排放总量 Total Volume of Industrial Waste Water Discharged		工业废水处理量 Industrial Waste Water Treated
		直排入海 Discharged Directly into the Sea	
天 津 Tianjin	18 691.93	43.40	29 160.99
唐 山 Tangshan	12 588.59	1 091.15	324 207.67
秦皇岛 Qinhuangdao	6 156.22	66.01	17 206.50
沧 州 Cangzhou	8 924.77	0.36	10 363.08
大 连 Dalian	26 153.78	19 616.45	7 719.78
丹 东 Dandong	4 780.18	7.76	4 660.55
锦 州 Jinzhou	4 940.49	80.31	4 233.02
营 口 Yingkou	2 789.93	804.27	3 679.36
盘 锦 Panjin	4 231.33	0.00	3 317.46
葫芦岛 Huludao	2 591.98	593.74	3 488.23
上 海 Shanghai	45 425.76	9 837.36	65 019.18
南 通 Nantong	14 584.29	0.00	13 356.70
连云港 Lianyungang	5 618.35	258.23	10 197.30
盐 城 Yancheng	17 555.17	442.02	20 170.21

8-5 续表1 continued

沿海城市 Coastal City	工业废水排放总量 Total Volume of Industrial Waste Water Discharged	直排入海 Discharged Directly into the Sea	工业废水处理量 Industrial Waste Water Treated
杭　州 Hangzhou	39 186.39	0.00	121 953.05
宁　波 Ningbo	19 666.24	6 035.81	30 005.43
温　州 Wenzhou	7 433.26	4.75	7 071.24
嘉　兴 Jiaxing	21 170.95	181.01	24 629.75
绍　兴 Shaoxing	27 244.73	0.00	26 222.56
舟　山 Zhoushan	2 093.76	1 277.51	939.56
台　州 Taizhou	6 278.31	171.43	5 712.45
福　州 Fuzhou	4 682.00	250.55	17 944.12
厦　门 Xiamen	27 260.82	22 465.67	28 668.71
莆　田 Putian	1 852.01	204.10	1 745.20
泉　州 Quanzhou	20 079.63	1 956.21	30 236.91
漳　州 Zhangzhou	25 484.57	17 594.99	13 992.59
宁　德 Ningde	1 447.40	15.89	1 545.23
青　岛 Qingdao	10 660.55	1 875.14	47 098.70
东　营 Dongying	10 110.55	0.00	28 454.27
烟　台 Yantai	9 529.75	2 729.83	16 074.97
潍　坊 Weifang	28 103.06	0.00	29 364.56
威　海 Weihai	2 740.17	102.44	3 254.73
日　照 Rizhao	7 911.05	4 171.29	8 539.71
滨　州 Binzhou	15 920.67	0.00	13 973.46

8-5 续表2 continued

沿海城市 Coastal City	工业废水排放总量 Total Volume of Industrial Waste Water Discharged	直排入海 Discharged Directly into the Sea	工业废水处理量 Industrial Waste Water Treated
广　州 Guangzhou	21 390.81	10.13	22 939.30
深　圳 Shenzhen	15 287.69	645.25	14 838.05
珠　海 Zhuhai	5 537.88	61.77	4 842.51
汕　头 Shantou	5 657.60	166.35	5 300.62
江　门 Jiangmen	11 749.74	91.64	13 875.34
湛　江 Zhanjiang	6 204.56	561.82	5 997.76
茂　名 Maoming	4 625.97	932.77	6 542.04
惠　州 Huizhou	8 320.34	705.81	8 071.56
汕　尾 Shanwei	1 954.17	753.65	577.41
阳　江 Yangjiang	2 166.05	1.00	1 302.27
东　莞 Dongguan	24 173.64	4 798.92	35 104.55
中　山 Zhongshan	8 913.83	0.00	9 736.13
潮　州 Chaozhou	2 982.17	0.00	1 375.19
揭　阳 Jieyang	6 074.25	0.00	5 595.74
北　海 Beihai	1 829.69	494.56	25 291.84
防城港 Fangchenggang	1 554.54	210.66	17 440.91
钦　州 Qinzhou	894.27	2.64	864.92
海　口 Haikou	824.88	0.00	816.15
三　亚 Sanya	69.38	2.64	48.77

8-6　沿海地带工业废水排放及处理情况
Discharge and Treatment of Industrial Waste Water by Coastal Counties

单位：万吨　(10 000 t)

地区 Region	工业废水排放总量 Total Volume of Industrial Waste Water Discharged	直排入海 Discharged Directly into the Sea	工业废水处理量 Industrial Waste Water Treated
天　津　Tianjin	**9 029.22**	**43.40**	**9 569.35**
河　北　Hebei	**6 300.83**	**22.51**	**89 774.92**
唐　山　Tangshan	1 579.73	21.14	77 740.45
秦皇岛　Qinhuangdao	3 825.99	1.01	11 196.57
沧　州　Cangzhou	895.11	0.36	837.90
辽　宁　Liaoning	**28 246.86**	**16 883.41**	**14 913.03**
大　连　Dalian	20 393.27	15 477.64	5 520.87
丹　东　Dandong	1 237.01	7.76	635.00
锦　州　Jinzhou	1 395.88	0.00	2 006.72
营　口　Yingkou	1 950.55	804.27	3 002.07
盘　锦　Panjin	817.32	0.00	666.53
葫芦岛　Huludao	2 452.83	593.74	3 081.84
上　海　Shanghai	**30 125.43**	**9 837.36**	**54 370.74**
江　苏　Jiangsu	**20 742.65**	**700.25**	**29 209.86**
南　通　Nantong	7 243.27	0.00	6 608.77
连云港　Lianyungang	3 801.19	258.23	8 632.83
盐　城　Yancheng	9 698.19	442.02	13 968.26

8-6 续表1 continued

地 区 Region	工业废水排放总量 Total Volume of Industrial Waste Water Discharged	直排入海 Discharged Directly into the Sea	工业废水处理量 Industrial Waste Water Treated
浙 江 Zhejiang	**69 028.73**	**7 460.65**	**68 061.36**
杭 州 Hangzhou	11 862.92	0.00	11 764.81
宁 波 Ningbo	17 515.76	6 005.64	19 783.67
温 州 Wenzhou	5 698.41	4.75	5 348.79
嘉 兴 Jiaxing	7 662.58	1.32	8 337.04
绍 兴 Shaoxing	19 755.32	0.00	17 566.51
舟 山 Zhoushan	2 093.76	1 277.51	939.56
台 州 Taizhou	4 439.98	171.43	4 320.98
福 建 Fujian	**73 321.63**	**42 466.88**	**74 272.54**
福 州 Fuzhou	3 571.42	250.55	15 549.46
厦 门 Xiamen	27 260.82	22 465.67	28 668.71
莆 田 Putian	1 828.24	183.57	1 696.15
泉 州 Quanzhou	18 173.73	1 956.21	21 464.55
漳 州 Zhangzhou	21 444.56	17 594.99	5 991.25
宁 德 Ningde	1 042.86	15.89	902.42
山 东 Shandong	**44 520.29**	**5 154.29**	**108 152.65**
青 岛 Qingdao	9 126.55	1 875.14	45 469.75
东 营 Dongying	9 660.87	0.00	28 084.96
烟 台 Yantai	7 976.56	2 729.83	14 653.55
潍 坊 Weifang	11 048.99	0.00	13 259.76
威 海 Weihai	2 039.41	100.72	2 543.98
日 照 Rizhao	1 194.58	448.60	1 682.73
滨 州 Binzhou	3 473.33	0.00	2 457.92

8-6 续表2 continued

地 区 Region	工业废水排放总量 Total Volume of Industrial Waste Water Discharged	直排入海 Discharged Directly into the Sea	工业废水处理量 Industrial Waste Water Treated
广 东 Guangdong	**87 859.23**	**2 136.61**	**102 902.90**
广 州 Guangzhou	10 715.10	10.13	12 421.77
深 圳 Shenzhen	12 881.20	584.10	13 173.99
珠 海 Zhuhai	2 556.99	36.35	2 127.58
汕 头 Shantou	5 609.56	166.35	5 294.93
江 门 Jiangmen	7 828.31	91.64	10 189.01
湛 江 Zhanjiang	6 073.23	541.25	5 861.85
茂 名 Maoming	1 893.50	9.28	3 783.60
惠 州 Huizhou	1 861.17	0.00	1 804.08
汕 尾 Shanwei	1 818.53	696.51	550.78
阳 江 Yangjiang	1 332.93	1.00	595.40
东 莞 Dongguan	24 173.64	0.00	35 104.55
中 山 Zhongshan	8 913.83	0.00	9 736.13
潮 州 Chaozhou	1 510.97	0.00	619.65
揭 阳 Jieyang	690.27	0.00	1 639.58
广 西 Guangxi	**5 134.26**	**2 739.83**	**39 385.10**
北 海 Beihai	1 829.69	494.56	25 291.84
防城港 Fangchenggang	1 052.22	210.66	11 299.92
钦 州 Qinzhou	2 252.35	2 034.61	2 793.34
海 南 Hainan	**804.29**	**0.00**	**682.42**
海 口 Haikou	734.91	0.00	633.65
三 亚 Sanya	69.38	0.00	48.77

注：1. 表中数据为沿海地带合计数（表8-9、表8-13同）。
2. 沿海地带是指有海岸线的县、县级市、区（包括直辖市和地级市的区）。

Note: 1. The data in the table are the total numbers for the coastal regions (The same for Tables 8-9, 8-13).
2.Coastal Zone refers to the counties, county-level cities and districts with coastlines (including the districts under the municipalities directly under the Central Government and the prefecture-level districts)

8-7 沿海地区一般工业固体废物倾倒丢弃、处理及综合利用情况
Discharge, Treatment and Multipurpose Utilization of Common Industrial Solid Wastes by Coastal Regions

单位：万吨 (10 000 t)

地区 Region		一般工业固体废物倾倒丢弃量 Volume of Common Industrial Solid Wastes Discharged	一般工业固体废物处置量 Volume of Common Industrial Solid Wastes Treated	一般工业固体废物综合利用量 Volume of Common Industrial Solid Wastes Comprehansively Utilized
总计 Total		**11.11**	**39 385.24**	**83 667.45**
环渤海经济区 Round-the-Bohai Sea Economic Zone	**合计 Total**	**9.08**	**35 515.18**	**48 815.31**
	辽宁 Liaoning	9.05	11 289.33	11 742.28
	河北 Hebei	0.00	23 428.61	18 356.16
	天津 Tianjin	0.00	9.67	1 582.44
	山东 Shandong	0.03	787.57	17 134.43
长江三角洲经济区 Yangtze River Delta Economic Zone	**合计 Total**	**0.04**	**521.41**	**16 588.34**
	江苏 Jiangsu	0.00	286.79	10 501.86
	上海 Shanghai	0.04	57.99	1 995.35
	浙江 Zhejiang	0.00	176.63	4 091.13
海峡西岸经济区 Economic Zone on the West Side of the Taiwan Straits	**合计 Total**	**0.05**	**962.29**	**7 543.88**
	福建 Fujian	0.05	962.29	7 543.88
珠江三角洲经济区 Zhujiang River Delta Economic Zone	**合计 Total**	**1.56**	**731.76**	**5 023.74**
	广东 Guangdong	1.56	731.76	5 023.74
环北部湾经济区 Round-the-Beibu Gulf Economic Zone	**合计 Total**	**0.38**	**1 654.60**	**5 696.18**
	广西 Guangxi	0.38	1 608.50	5 424.94
	海南 Hainan	0.00	46.10	271.24

8-8 沿海城市一般工业固体废物倾倒丢弃、处理及综合利用情况
Discharge, Treatment and Multipurposed Utilization of Common Industrial Solid Wastes by Coastal Cities

单位：万吨 (10 000 t)

沿海城市 Coastal City	一般工业固体废物 倾倒丢弃量 Volume of Common Industrial Solid Wastes Discharged	一般工业固体废物 处置量 Volume of Common Industrial Solid Wastes Treated	一般工业固体废物 综合利用量 Volume of Common Industrial Solid Wastes Comprehensively Utilized
天　津　Tianjin	0.00	9.67	1 582.44
唐　山　Tangshan	0.00	2 571.93	7 611.14
秦皇岛　Qinhuangdao	0.00	693.97	868.41
沧　州　Cangzhou	0.00	1.70	402.21
大　连　Dalian	0.00	49.93	467.91
丹　东　Dandong	0.00	367.55	243.54
锦　州　Jinzhou	0.00	2.44	235.18
营　口　Yingkou	0.00	0.76	568.12
盘　锦　Panjin	0.00	10.71	125.76
葫芦岛　Huludao	1.04	6.13	362.82
上　海　Shanghai	0.04	57.99	1 995.35
南　通　Nantong	0.00	8.35	429.61
连云港　Lianyungang	0.00	1.12	478.36
盐　城　Yancheng	0.00	58.84	230.06

8-8 续表1 continued

沿海城市 Coastal City	一般工业固体废物倾倒丢弃量 Volume of Common Industrial Solid Wastes Discharged	一般工业固体废物处置量 Volume of Common Industrial Solid Wastes Treated	一般工业固体废物综合利用量 Volume of Common Industrial Solid Wastes Comprehansively Utilized
杭　州 Hangzhou	0.00	37.71	647.92
宁　波 Ningbo	0.00	48.66	1 167.91
温　州 Wenzhou	0.00	2.15	211.52
嘉　兴 Jiaxing	0.00	21.97	466.74
绍　兴 Shaoxing	0.00	22.64	333.84
舟　山 Zhoushan	0.00	0.15	74.44
台　州 Taizhou	0.00	11.60	314.87
福　州 Fuzhou	0.00	43.60	763.65
厦　门 Xiamen	0.00	8.52	96.97
莆　田 Putian	0.00	0.36	76.54
泉　州 Quanzhou	0.00	15.10	775.43
漳　州 Zhangzhou	0.00	11.03	197.41
宁　德 Ningde	0.00	11.80	230.34
青　岛 Qingdao	0.00	22.44	791.15
东　营 Dongying	0.00	4.90	321.75
烟　台 Yantai	0.00	369.79	2 083.25
潍　坊 Weifang	0.03	22.39	812.20
威　海 Weihai	0.00	21.20	323.98
日　照 Rizhao	0.00	6.49	956.76
滨　州 Binzhou	0.00	11.81	664.09

8-8 续表2 continued

沿海城市 Coastal City	一般工业固体废物倾倒丢弃量 Volume of Common Industrial Solid Wastes Discharged	一般工业固体废物处置量 Volume of Common Industrial Solid Wastes Treated	一般工业固体废物综合利用量 Volume of Common Industrial Solid Wastes Comprehansively Utilized
广 州 Guangzhou	0.00	21.93	528.88
深 圳 Shenzhen	0.00	46.23	63.22
珠 海 Zhuhai	0.00	17.02	245.67
汕 头 Shantou	0.01	1.54	155.47
江 门 Jiangmen	0.00	23.32	234.38
湛 江 Zhanjiang	0.07	4.55	257.74
茂 名 Maoming	0.00	4.49	104.05
惠 州 Huizhou	0.00	41.49	116.03
汕 尾 Shanwei	0.00	0.27	69.33
阳 江 Yangjiang	0.20	0.09	235.29
东 莞 Dongguan	0.66	109.20	417.50
中 山 Zhongshan	0.46	27.43	71.94
潮 州 Chaozhou	0.00	0.11	79.91
揭 阳 Jieyang	0.08	0.06	115.22
北 海 Beihai	0.00	0.32	228.55
防城港 Fangchenggang	0.00	0.77	145.46
钦 州 Qinzhou	0.01	1.24	100.45
海 口 Haikou	0.00	0.41	6.00
三 亚 Sanya	0.00	0.00	5.18

8-9 沿海地带一般工业固体废物倾倒丢弃、处理及综合利用情况
Discharge, Treatment and Multipurposed Utilization of Common Industrial Solid Wastes by Coastal Counties

单位：万吨　　(10 000 t)

地　区 Region	一般工业固体废物倾倒丢弃量 Volume of Common Industrial Solid Wastes Discharged	一般工业固体废物处置量 Volume of Common Industrial Solid Wastes Treated	一般工业固体废物综合利用量 Volume of Common Industrial Solid Wastes Comprehansively Utilized
天　津　Tianjin	**0.00**	**7.29**	**393.13**
河　北　Hebei	**0.00**	**108.98**	**1 729.91**
唐　山　Tangshan	0.00	4.97	1 254.04
秦皇岛　Qinhuangdao	0.00	104.01	473.62
沧　州　Cangzhou	0.00	0.00	2.25
辽　宁　Liaoning	**0.00**	**80.06**	**1 175.63**
大　连　Dalian	0.00	46.22	408.39
丹　东　Dandong	0.00	27.17	14.12
锦　州　Jinzhou	0.00	0.09	14.25
营　口　Yingkou	0.00	0.53	516.23
盘　锦　Panjin	0.00	0.00	12.29
葫芦岛　Huludao	0.00	6.05	210.35
上　海　Shanghai	**0.04**	**32.60**	**1 747.45**
江　苏　Jiangsu	**0.00**	**61.67**	**717.29**
南　通　Nantong	0.00	4.67	217.30
连云港　Lianyungang	0.00	0.69	349.20
盐　城　Yancheng	0.00	56.31	150.79

8-9 续表1 continued

地 区 Region	一般工业固体废物倾倒丢弃量 Volume of Common Industrial Solid Wastes Discharged	一般工业固体废物处置量 Volume of Common Industrial Solid Wastes Treated	一般工业固体废物综合利用量 Volume of Common Industrial Solid Wastes Comprehensively Utilized
浙 江 Zhejiang	**0.00**	**81.29**	**2 157.77**
杭 州 Hangzhou	0.00	16.02	155.88
宁 波 Ningbo	0.00	46.23	1 137.27
温 州 Wenzhou	0.00	1.11	176.57
嘉 兴 Jiaxing	0.00	0.53	134.20
绍 兴 Shaoxing	0.00	5.97	188.67
舟 山 Zhoushan	0.00	0.15	74.44
台 州 Taizhou	0.00	11.28	290.74
福 建 Fujian	**0.00**	**59.80**	**1 459.45**
福 州 Fuzhou	0.00	33.04	680.71
厦 门 Xiamen	0.00	8.52	96.97
莆 田 Putian	0.00	0.18	44.32
泉 州 Quanzhou	0.00	9.60	300.31
漳 州 Zhangzhou	0.00	8.36	119.84
宁 德 Ningde	0.00	0.10	217.30
山 东 Shandong	**0.00**	**415.76**	**4 276.07**
青 岛 Qingdao	0.00	21.70	586.97
东 营 Dongying	0.00	3.21	278.66
烟 台 Yantai	0.00	366.34	2 025.44
潍 坊 Weifang	0.00	2.83	225.47
威 海 Weihai	0.00	20.76	134.66
日 照 Rizhao	0.00	0.27	776.28
滨 州 Binzhou	0.00	0.65	248.59

8-9 续表2 continued

地 区 Region	一般工业固体废物倾倒丢弃量 Volume of Common Industrial Solid Wastes Discharged	一般工业固体废物处置量 Volume of Common Industrial Solid Wastes Treated	一般工业固体废物综合利用量 Volume of Common Industrial Solid Wastes Comprehansively Utilized
广 东 Guangdong	**1.30**	**229.73**	**2 083.15**
广 州 Guangzhou	0.00	17.48	441.41
深 圳 Shenzhen	0.00	45.14	62.81
珠 海 Zhuhai	0.00	1.51	8.93
汕 头 Shantou	0.01	1.54	155.47
江 门 Jiangmen	0.00	18.07	206.62
湛 江 Zhanjiang	0.01	4.47	256.23
茂 名 Maoming	0.00	4.39	69.50
惠 州 Huizhou	0.00	0.08	68.54
汕 尾 Shanwei	0.00	0.26	69.15
阳 江 Yangjiang	0.16	0.03	72.99
东 莞 Dongguan	0.66	109.20	417.50
中 山 Zhongshan	0.46	27.43	71.94
潮 州 Chaozhou	0.00	0.07	76.49
揭 阳 Jieyang	0.00	0.06	105.57
广 西 Guangxi	**0.00**	**1.31**	**364.08**
北 海 Beihai	0.00	0.32	228.55
防城港 Fangchenggang	0.00	0.76	69.81
钦 州 Qinzhou	0.00	0.23	65.72
海 南 Hainan	**0.00**	**0.41**	**8.37**
海 口 Haikou	0.00	0.41	3.19
三 亚 Sanya	0.00	0.00	5.18

8-10 主要沿海城市工业废气排放及处理情况
Emission and Treatment of Industrial Waste Gas in Major Coastal Cities

城 市 City	工业废气排放量（亿立方米） Industrial Waste Gas Emission (100 000 000 m^3)	工业二氧化硫排放量（吨） Industrial Sulphur Dioxide Emission (t)	工业二氧化硫处理量（吨） Industrial Sulphur Dioxide Treated (t)	工业氮氧化物排放量（吨） Industrial Nitrogen Oxide Emission (t)	工业氮氧化物处理量（吨） Industrial Nitrogen Oxide Treated (t)
天 津 Tianjin	8 080.0	207 792.8	1 218 063.4	250 645.6	99 285.1
唐 山 Tangshan	33 109.6	282 806.0	431 934.2	282 545.3	33 200.6
秦皇岛 Qinhuangdao	3 021.7	72 501.1	78 929.3	53 459.3	10 881.7
沧 州 Cangzhou	2 345.0	40 688.5	122 456.9	48 316.2	13 906.6
大 连 Dalian	2 261.7	102 937.8	139 925.6	101 136.0	22 098.2
丹 东 Dandong	872.3	30 042.5	48 007.2	20 545.2	6 337.2
锦 州 Jinzhou	787.1	36 584.5	45 504.7	18 221.1	6 238.2
营 口 Yingkou	3 507.7	52 069.2	45 466.8	48 326.5	5 470.3
盘 锦 Panjin	1 036.3	53 562.6	35 176.3	25 875.7	0.0
葫芦岛 Huludao	1 557.9	72 920.1	463 808.4	49 557.0	16 549.2
上 海 Shanghai	13 344.1	172 867.4	365 972.8	262 345.6	66 483.6
南 通 Nantong	2 213.5	63 010.5	152 364.3	51 706.1	16 060.1
连云港 Lianyungang	964.6	43 461.7	62 669.2	29 657.2	6 950.2
盐 城 Yancheng	1 304.6	44 815.0	89 206.1	30 785.5	12 682.6
杭 州 Hangzhou	4 757.8	82 020.6	89 336.7	67 282.8	22 384.1
宁 波 Ningbo	6 487.0	134 629.6	442 120.5	212 519.4	77 449.7
温 州 Wenzhou	1 673.7	34 479.1	82 270.2	34 600.5	21 171.7

8-10 续表1 continued

城市 City	工业废气排放量（亿立方米）Industrial Waste Gas Emission ($100\ 000\ 000\ m^3$)	工业二氧化硫排放量（吨）Industrial Sulphur Dioxide Emission (t)	工业二氧化硫处理量（吨）Industrial Sulphur Dioxide Treated (t)	工业氮氧化物排放量（吨）Industrial Nitrogen Oxide Emission (t)	工业氮氧化物处理量（吨）Industrial Nitrogen Oxide Treated (t)
嘉兴 Jiaxing	2 055.0	72 959.9	127 738.0	43 744.2	20 709.9
绍兴 Shaoxing	1 420.3	59 635.1	59 465.6	41 781.4	5 622.5
舟山 Zhoushan	196.1	13 687.0	14 515.5	15 595.4	0.4
台州 Taizhou	1 507.2	37 571.1	151 468.8	29 140.7	22 427.0
福州 Fuzhou	3 551.5	76 043.4	164 390.0	72 284.0	46 966.6
厦门 Xiamen	1 119.5	18 772.3	34 825.3	11 907.9	15 112.8
莆田 Putian	584.3	11 227.4	51 429.7	25 090.2	16 386.5
泉州 Quanzhou	3 254.0	90 614.7	145 518.7	70 110.5	9 366.8
漳州 Zhangzhou	1 400.9	36 156.2	96 499.4	41 249.0	25 776.7
宁德 Ningde	472.3	14 944.7	47 656.1	14 874.0	35 909.6
青岛 Qingdao	2 128.7	69 336.8	194 654.6	64 938.4	64 904.1
东营 Dongying	1 102.4	52 817.5	311 842.5	31 006.8	5 809.1
烟台 Yantai	3 150.9	79 834.3	367 877.3	83 364.6	7 008.1
潍坊 Weifang	2 608.3	128 226.6	253 192.8	76 368.3	4 965.3
威海 Weihai	884.6	36 211.6	82 669.6	30 976.8	6 406.9
日照 Rizhao	4 394.1	52 084.2	75 542.7	55 471.9	11 804.6
滨州 Binzhou	1 972.3	80 330.2	125 058.8	33 722.8	6 359.1
广州 Guangzhou	3 774.0	65 589.5	381 682.9	57 164.4	49 691.9

8-10 续表2 continued

城 市 City	工业废气排放量（亿立方米） Industrial Waste Gas Emission (100 000 000 m^3)	工业二氧化硫排放量（吨） Industrial Sulphur Dioxide Emission (t)	工业二氧化硫处理量（吨） Industrial Sulphur Dioxide Treated (t)	工业氮氧化物排放量（吨） Industrial Nitrogen Oxide Emission (t)	工业氮氧化物处理量（吨） Industrial Nitrogen Oxide Treated (t)
深 圳 Shenzhen	2 067.3	5 713.8	51 795.7	18 402.7	9 563.3
珠 海 Zhuhai	1 327.6	22 653.4	67 027.1	40 932.7	7 742.9
汕 头 Shantou	927.4	29 059.9	77 909.8	24 378.4	5 015.6
江 门 Jiangmen	1 482.7	60 068.6	116 372.2	51 044.2	21 433.5
湛 江 Zhanjiang	1 093.8	22 877.0	96 150.7	17 043.5	11 510.2
茂 名 Maoming	897.0	30 105.8	409 412.4	17 238.0	309.4
惠 州 Huizhou	1 509.5	30 029.5	52 107.2	34 757.5	24 000.8
汕 尾 Shanwei	493.2	11 793.9	53 759.8	11 871.3	3 867.8
阳 江 Yangjiang	796.7	22 207.9	34 349.5	12 877.5	6 327.9
东 莞 Dongguan	3 047.3	113 896.5	137 685.0	92 295.7	24 017.6
中 山 Zhongshan	713.2	22 492.2	10 020.5	18 337.6	0.0
潮 州 Chaozhou	924.9	13 444.2	70 749.3	16 428.4	9 235.2
揭 阳 Jieyang	622.1	17 397.8	72 087.4	11 829.7	6 673.2
北 海 Beihai	587.8	11 912.0	28 598.9	14 135.9	1 413.6
防城港 Fangchenggang	1 050.3	20 245.7	35 913.7	20 410.6	534.5
钦 州 Qinzhou	530.5	16 178.7	34 769.7	15 138.0	2.1
海 口 Haikou	50.0	1 797.8	0.0	85.9	0.0
三 亚 Sanya	57.5	2.6	0.0	123.1	0.0

8-10 续表3 continued

城 市 City	工业烟（粉）尘排放量（亿立方米） Industrial Soot (Dust) Emission (100 000 000 m^3)	工业烟粉尘处理量（吨） Industrial Soot (Dust) Treated (t)	生活二氧化硫排放量（吨） Household Sulphur Dioxide Emission (t)	生活氮氧化物排放量（吨） Household Nitrogen Oxide Emission (t)	生活烟尘排放量（吨） Household Soot Emission (t)
天 津 Tianjin	62 766.3	6 640 409.5	8 959.0	5 220.8	18 400.0
唐 山 Tangshan	478 573.8	15 879 442.8	9 446.2	4 410.0	1 090.0
秦皇岛 Qinhuangdao	78 091.9	2 079 643.9	7 886.5	2 367.4	5 640.7
沧 州 Cangzhou	54 621.4	1 742 599.4	9 228.6	1 248.2	7 340.0
大 连 Dalian	46 331.8	4 179 954.0	16 365.9	2 403.0	12 015.0
丹 东 Dandong	28 656.2	660 686.5	4 194.6	1 069.1	1 600.0
锦 州 Jinzhou	20 004.6	838 189.0	5 116.1	928.8	5 160.0
营 口 Yingkou	37 292.3	1 921 276.5	3 857.5	774.9	3 123.6
盘 锦 Panjin	12 703.0	684 955.3	5 002.4	794.3	3 908.0
葫芦岛 Huludao	19 552.1	1 500 232.4	2 210.2	346.1	1 625.0
上 海 Shanghai	67 173.9	6 540 983.5	42 947.0	23 474.0	6 451.0
南 通 Nantong	33 970.4	2 279 475.0	2 121.0	240.0	978.0
连云港 Lianyungang	19 179.0	1 834 955.0	6 273.9	811.8	3 026.3
盐 城 Yancheng	29 705.2	1 134 283.5	2 308.0	336.2	1 281.1
杭 州 Hangzhou	40 242.7	4 172 828.6	632.8	335.4	135.5
宁 波 Ningbo	25 274.9	6 453 740.8	2 130.8	545.0	3 406.0
温 州 Wenzhou	18 782.6	1 084 489.6	668.6	423.3	467.5

8-10 续表4 continued

城 市 City	工业烟（粉）尘排放量（亿立方米） Industrial Soot (Dust) Emission (100 000 000 m^3)	工业烟粉尘处理量（吨） Industrial Soot (Dust) Treated (t)	生活二氧化硫排放量（吨） Household Sulphur Dioxide Emission (t)	生活氮氧化物排放量（吨） Household Nitrogen Oxide Emission (t)	生活烟尘排放量（吨） Household Soot Emission (t)
嘉 兴 Jiaxing	24 752.4	2 114 643.3	385.0	93.0	45.0
绍 兴 Shaoxing	30 943.9	1 333 616.1	869.9	199.0	243.7
舟 山 Zhoushan	4 558.3	114 444.4	288.0	98.3	180.6
台 州 Taizhou	12 297.2	1 011 336.9	788.6	136.4	92.8
福 州 Fuzhou	43 483.3	2 918 548.9	1 279.0	169.0	547.0
厦 门 Xiamen	3 986.7	463 006.2	486.0	152.0	190.0
莆 田 Putian	4 557.5	216 175.0	1 400.8	388.0	800.0
泉 州 Quanzhou	52 626.1	1 544 949.0	3 199.4	558.6	1 771.3
漳 州 Zhangzhou	15 079.6	928 513.6	779.3	91.7	458.4
宁 德 Ningde	25 060.5	560 603.2	1 193.0	70.0	534.4
青 岛 Qingdao	27 802.7	2 573 399.6	27 497.5	2 682.7	9 953.8
东 营 Dongying	5 636.2	1 338 460.6	1 720.9	864.7	925.5
烟 台 Yantai	34 945.4	4 994 524.7	8 053.1	540.2	5 608.1
潍 坊 Weifang	33 823.4	4 894 349.9	12 480.2	1 315.9	6 904.8
威 海 Weihai	7 998.9	1 185 624.3	8 370.6	639.4	8 901.6
日 照 Rizhao	29 328.3	3 561 092.1	13 152.8	1 206.4	5 300.0
滨 州 Binzhou	21 841.6	2 480 607.0	3 642.6	555.6	2 465.1
广 州 Guangzhou	16 659.5	3 333 042.2	663.0	276.0	214.0

8-10 续表5 continued

城 市 City	工业烟（粉）尘排放量（亿立方米） Industrial Soot (Dust) Emission (100 000 000 m^3)	工业烟粉尘处理量（吨） Industrial Soot (Dust) Treated (t)	生活二氧化硫排放量（吨） Household Sulphur Dioxide Emission (t)	生活氮氧化物排放量（吨） Household Nitrogen Oxide Emission (t)	生活烟尘排放量（吨） Household Soot Emission (t)
深 圳 Shenzhen	1 927.1	387 689.0	238.0	885.0	0.0
珠 海 Zhuhai	9 595.2	822 876.2	21.4	44.4	3.6
汕 头 Shantou	7 358.6	853 453.6	86.2	117.0	55.2
江 门 Jiangmen	12 510.1	1 419 522.8	110.0	4.2	70.0
湛 江 Zhanjiang	9 892.5	1 014 204.0	2 580.2	402.8	263.1
茂 名 Maoming	11 489.0	689 290.2	2 124.3	607.1	2 041.8
惠 州 Huizhou	23 017.3	1 168 300.4	266.2	313.3	70.1
汕 尾 Shanwei	4 134.7	490 997.3	659.0	145.2	475.1
阳 江 Yangjiang	12 141.9	572 721.1	1 210.0	139.0	474.0
东 莞 Dongguan	15 543.1	621 988.5	2 161.0	156.0	450.0
中 山 Zhongshan	17 412.4	309 488.0	238.0	273.4	200.0
潮 州 Chaozhou	6 128.3	938 276.1	769.3	120.2	90.2
揭 阳 Jieyang	2 962.3	243 959.7	4 245.9	624.4	2 017.5
北 海 Beihai	5 754.5	541 720.4	1 240.3	140.7	395.2
防城港 Fangchenggang	12 198.3	1 111 943.4	420.0	52.7	162.5
钦 州 Qinzhou	5 844.2	281 591.4	1 117.0	76.7	237.3
海 口 Haikou	1 149.3	541.5	11.4	17.3	4.5
三 亚 Sanya	219.9	4 972.2	51.0	12.4	16.0

8-11 沿海地区污染治理项目情况
Pollution Treatment Projects in Coastal Regions

单位：个 (unit)

地区 Region		当年安排施工项目 Arranged for Construction in the Year		当年竣工项目 Completed in the Year	
		治理废水 Treatment of Waste Water	治理固体废物 Treatment of Solid Wastes	治理废水 Treatment of Waste Water	治理固体废物 Treatment of Solid Wastes
总 计 Total		**964**	**151**	**1 003**	**152**
环渤海经济区 Round-the-Bohai Sea Economic Zone	**合 计 Total**	**144**	**66**	**182**	**67**
	辽 宁 Liaoning	3	8	13	10
	河 北 Hebei	41	52	50	52
	天 津 Tianjin	14	0	13	0
	山 东 Shandong	86	6	106	5
长江三角洲经济区 Yangtze River Delta Economic Zone	**合 计 Total**	**531**	**12**	**488**	**14**
	江 苏 Jiangsu	169	1	174	3
	上 海 Shanghai	30	2	28	3
	浙 江 Zhejiang	332	9	286	8
海峡西岸经济区 Economic Zone on the West Side of the Taiwan Straits	**合 计 Total**	**83**	**50**	**94**	**48**
	福 建 Fujian	83	50	94	48
珠江三角洲经济区 Zhujiang River Delta Economic Zone	**合 计 Total**	**152**	**17**	**184**	**18**
	广 东 Guangdong	152	17	184	18
环北部湾经济区 Round-the-Beibu Gulf Economic Zone	**合 计 Total**	**54**	**6**	**55**	**5**
	广 西 Guangxi	52	6	53	5
	海 南 Hainan	2	0	2	0

8-12 沿海城市污染治理项目情况
Pollution Treatment Projects in Coastal Cities

单位：个 (unit)

沿海城市 Coastal City	当年安排施工项目 Arranged for Construction in the Year		当年竣工项目 Completed in the Year	
	治理废水 Treatment of Waste Water	治理固体废物 Treatment of Solid Wastes	治理废水 Treament of Waste Water	治理固体废物 Treatment of Solid Wastes
天　津 Tianjin	14	0	13	0
唐　山 Tangshan	3	0	3	0
秦皇岛 Qinhuangdao	0	0	0	0
沧　州 Cangzhou	3	0	5	0
大　连 Dalian	0	1	1	1
丹　东 Dandong	1	0	1	0
锦　州 Jinzhou	0	0	3	0
营　口 Yingkou	0	0	0	0
盘　锦 Panjin	0	5	0	6
葫芦岛 Huludao	1	0	1	0
上　海 Shanghai	30	2	28	3
南　通 Nantong	53	0	48	2
连云港 Lianyungang	18	0	18	0
盐　城 Yancheng	0	0	0	0

8-12 续表1 continued

沿海城市 Coastal City	当年安排施工项目 Arranged for Construction in the Year		当年竣工项目 Completed in the Year	
	治理废水 Treatment of Waste Water	治理固体废物 Treatment of Solid Wastes	治理废水 Treament of Waste Water	治理固体废物 Treatment of Solid Wastes
杭　州 Hangzhou	39	1	40	1
宁　波 Ningbo	23	0	18	0
温　州 Wenzhou	0	0	2	0
嘉　兴 Jiaxing	35	0	32	0
绍　兴 Shaoxing	69	2	71	2
舟　山 Zhoushan	2	0	1	0
台　州 Taizhou	29	3	21	2
福　州 Fuzhou	2	0	2	0
厦　门 Xiamen	8	0	23	0
莆　田 Putian	0	1	0	0
泉　州 Quanzhou	26	42	6	41
漳　州 Zhangzhou	12	2	19	2
宁　德 Ningde	2	0	3	0
青　岛 Qingdao	3	1	3	0
东　营 Dongying	5	0	8	0
烟　台 Yantai	9	2	12	2
潍　坊 Weifang	18	0	17	0
威　海 Weihai	9	0	4	0
日　照 Rizhao	1	0	4	0
滨　州 Binzhou	2	0	3	0

8-12 续表2 continued

沿海城市 Coastal City	当年安排施工项目 Arranged for Construction in the Year		当年竣工项目 Completed in the Year	
	治理废水 Treatment of Waste Water	治理固体废物 Treatment of Solid Wastes	治理废水 Treament of Waste Water	治理固体废物 Treatment of Solid Wastes
广　州 Guangzhou	7	0	11	0
深　圳 Shenzhen	18	1	21	1
珠　海 Zhuhai	6	0	9	0
汕　头 Shantou	7	0	8	0
江　门 Jiangmen	21	1	21	1
湛　江 Zhanjiang	15	0	17	0
茂　名 Maoming	1	0	3	1
惠　州 Huizhou	5	0	5	0
汕　尾 Shanwei	0	0	0	0
阳　江 Yangjiang	3	0	5	0
东　莞 Dongguan	23	1	29	1
中　山 Zhongshan	0	0	0	0
潮　州 Chaozhou	1	0	1	0
揭　阳 Jieyang	0	0	0	0
北　海 Beihai	4	0	5	0
防城港 Fangchenggang	2	0	2	0
钦　州 Qinzhou	3	0	3	1
海　口 Haikou	0	0	0	0
三　亚 Sanya	0	0	0	0

8-13 沿海地带污染治理项目情况
Pollution Treatment Projects in Coastal Counties

单位：个 (unit)

地 区 Region	当年安排施工项目 Arranged for Construction in the Year		当年竣工项目 Completed in the Year	
	治理废水 Treatment of Waste Water	治理固体废物 Treatment of Solid Wastes	治理废水 Treament of Waste Water	治理固体废物 Treatment of Solid Wastes
天 津 Tianjin	**1**	**0**	**3**	**0**
河 北 Hebei	**2**	**0**	**2**	**0**
唐 山 Tangshan	2	0	1	0
秦皇岛 Qinhuangdao	0	0	0	0
沧 州 Cangzhou	0	0	1	0
辽 宁 Liaoning	**1**	**1**	**3**	**1**
大 连 Dalian	0	1	1	1
丹 东 Dandong	0	0	0	0
锦 州 Jinzhou	0	0	1	0
营 口 Yingkou	0	0	0	0
盘 锦 Panjin	0	0	0	0
葫芦岛 Huludao	1	0	1	0
上 海 Shanghai	**18**	**2**	**17**	**3**
江 苏 Jiangsu	**59**	**0**	**54**	**2**
南 通 Nantong	44	0	41	2
连云港 Lianyungang	15	0	13	0
盐 城 Yancheng	0	0	0	0

8-13 续表1 continued

地 区 Region	当年安排施工项目 Arranged for Construction in the Year		当年竣工项目 Completed in the Year	
	治理废水 Treatment of Waste Water	治理固体废物 Treatment of Solid Wastes	治理废水 Treament of Waste Water	治理固体废物 Treatment of Solid Wastes
浙 江 Zhejiang	**125**	**5**	**101**	**4**
杭 州 Hangzhou	14	0	14	0
宁 波 Ningbo	22	0	18	0
温 州 Wenzhou	0	0	2	0
嘉 兴 Jiaxing	19	0	16	0
绍 兴 Shaoxing	40	2	31	2
舟 山 Zhoushan	2	0	1	0
台 州 Taizhou	28	3	19	2
福 建 Fujian	**39**	**9**	**38**	**7**
福 州 Fuzhou	0	0	0	0
厦 门 Xiamen	8	0	23	0
莆 田 Putian	0	1	0	0
泉 州 Quanzhou	24	6	4	5
漳 州 Zhangzhou	6	2	9	2
宁 德 Ningde	1	0	2	0
山 东 Shandong	**34**	**3**	**40**	**2**
青 岛 Qingdao	3	1	3	0
东 营 Dongying	5	0	8	0
烟 台 Yantai	8	2	11	2
潍 坊 Weifang	8	0	10	0
威 海 Weihai	9	0	4	0
日 照 Rizhao	1	0	3	0
滨 州 Binzhou	0	0	1	0

8-13 续表2 continued

地 区 Region	当年安排施工项目 Arranged for Construction in the Year		当年竣工项目 Completed in the Year	
	治理废水 Treatment of Waste Water	治理固体废物 Treatment of Solid Wastes	治理废水 Treament of Waste Water	治理固体废物 Treatment of Solid Wastes
广 东 Guangdong	**70**	**1**	**75**	**1**
广 州 Guangzhou	6	0	9	0
深 圳 Shenzhen	18	1	20	1
珠 海 Zhuhai	3	0	3	0
汕 头 Shantou	7	0	8	0
江 门 Jiangmen	12	0	5	0
湛 江 Zhanjiang	15	0	17	0
茂 名 Maoming	1	0	2	0
惠 州 Huizhou	5	0	5	0
汕 尾 Shanwei	0	0	0	0
阳 江 Yangjiang	3	0	5	0
东 莞 Dongguan	0	0	0	0
中 山 Zhongshan	0	0	0	0
潮 州 Chaozhou	0	0	1	0
揭 阳 Jieyang	0	0	0	0
广 西 Guangxi	**5**	**0**	**7**	**0**
北 海 Beihai	4	0	5	0
防城港 Fangchenggang	0	0	0	0
钦 州 Qinzhou	1	0	2	0
海 南 Hainan	**0**	**0**	**0**	**0**
海 口 Haikou	0	0	0	0
三 亚 Sanya	0	0	0	0

8-14 沿海区域海洋类型自然保护区建设情况
Construction of Marine-Type Nature Reserves in Coastal Regions

地　区 Region	保护区数量（个） Number of Nature Reserves (unit)	按保护级别分（个） By Level of Protection (unit)		按保护类型分（个） By Type of Protection (unit)				保护区面积（平方千米） Area(km^2)
		国家级 National	地方级 Provincial	海洋和海岸生态系统 Marine and Coastal Ecosystem	海洋自然历史遗迹 Marine Natural and Historical Relics	海洋生物多样性 Marine Biodiversity	其他 Others	
合　计 Total	**133**	**30**	**103**	**48**	**7**	**71**	**7**	**48 503**
环渤海经济区 Round-the-Bohai Sea Economic Zone	37	11	26	19	3	15		16 101
长江三角洲经济区 Yangtze River Delta Economic Zone	12	5	7	2	2	4	4	2 386
海峡西岸经济区 Economic Zone on the West Side of the Taiwan Straits	12	3	9	5	2	5	0	1 089
珠江三角洲经济区 Zhujiang River Delta Economic Zone	50	5	45	20	0	28	2	3 820
环北部湾经济区 Round-the-Beibu Gulf Economic Zone	22	6	16	2	0	19	1	25 107

8-15 沿海地区海洋类型自然保护区建设情况
Construction of Marine-Type Nature Reserves

地 区 Region	保护区数量（个） Number of Nature Reserves (unit)	按保护级别分（个） By Level of Protection (unit)		按保护类型分（个） By Type of Protection (unit)				保护区面积（平方千米） Area (km^2)
		国家级 National	地方级 Provincial	海洋和海岸生态系统 Marine and Coastal Ecosystem	海洋自然历史遗迹 Marine Natural and Historical Relics	海洋生物多样性 Marine Biodiversity	其他 Others	
合 计 Total	**133**	**30**	**103**	**48**	**7**	**71**	**7**	**48 503**
天 津 Tianjin	1	1		1				359
河 北 Hebei	3	1	2	1	1	1		339
辽 宁 Liaoning	15	5	10	10	2	3		9 860
上 海 Shanghai	4	2	2	1			3	941
江 苏 Jiangsu	4	1	3			4		724
浙 江 Zhejiang	4	2	2	1	2		1	721
福 建 Fujian	12	3	9	5	2	5		1 089
山 东 Shandong	18	4	14	7		11		5 543
广 东 Guangdong	50	5	45	20		28	2	3 820
广 西 Guangxi	2	2		2				110
海 南 Hainan	20	4	16			19	1	24 997

8-16 全国海洋生态监控区基本情况
Basic Condition of the Marine Ecological Monitoring Areas throughout the Country

生态监控区 Ecological Monitoring Area	所在地 Location	面积（平方千米）Area (km^2)	主要生态系统类型 Major Types of Ecosystem	多样性指数* Diversity Indices		
				浮游植物 Phyto-plankton	大型浮游动物 Macrozoo-plankton	底栖生物 Macrobendthos
双台子河口 Shungtaizi Estuary	辽宁省 Liaoning Province	3 000	河口 Estuary	2.45	2.00	
锦州湾 Jinzhou Bay	辽宁省 Liaoning Province	650	海湾 Bay	2.08	1.11	
滦河口-北戴河 Luanhe River Mouth-Beidaihe	河北省 Hebei Province	900	河口 Estuary	2.13	1.36	1.62
渤海湾 Bohai Bay	天津市 Tianjin Municipality	3 000	海湾 Bay	2.00	1.73	2.49
莱州湾 Laizhou Bay	山东省 Shandong Province	3 770	海湾 Bay	1.64	2.52	2.74
黄河口 Yellow River Mouth	山东省 Shandong Province	2 600	河口 Estuary	1.94	1.46	1.80
苏北浅滩 North Jiangsu Bank	江苏省 Jiangsu Province	15 400	滩涂湿地 Mudflat and Wetland	2.58	2.02	1.62
长江口 Yangtze River Mouth	上海市 Shanghai Municipality	13 668	河口 Estuary	2.18	1.97	1.70
杭州湾 Hangzhou Bay	上海市 浙江省 Shanghai Municipality Zhejiang Province	5 000	海湾 Bay	1.80	2.42	0.00
乐清湾 Yueqing Bay	浙江省 Zhejiang Province	464	海湾 Bay	2.22	2.52	0.85

8-16 续表 continued

生态监控区 Ecological Monitoring Area	所在地 Location	面积（平方千米） Area (km^2)	主要生态系统类型 Major Types of Ecosystem	多样性指数* Diversity Indices		
				浮游植物 Phyto-plankton	大型浮游动物 Macrozoo-plankton	底栖生物 Macrobendthos
闽东沿岸	福建省	5 063	海 湾	1.32	2.68	2.59
Coastal East Fujian	Fujian Province		Bay			
大亚湾	广东省	1 200	海 湾	1.66	2.99	2.00
Daya Bay	Guangdong Province		Bay			
珠江口	广东省	3 980	河 口	1.65	2.29	1.57
Zhujiang River Mouth	Guangdong Province		Estuary			
雷州半岛西南沿岸	广东省	1 150	珊瑚礁			
Southwest Coast of Leizhou Peninsula	Guangdong Province		Coral Reef			
广西北海	广西壮族自治区	120	珊瑚礁 红树林 海草床			
Beihai, Guangxi	Guangxi Zhuang Nationality Autonomous Region		Coral Reef Mangroves Seagrass Bed			
北仑河口	广西壮族自治区	150	红树林			
Beilun River Mouth	Guangxi Zhuang Nationality Autonomous Region		Mangroves			
海南东海岸	海南省	3 750	珊瑚礁 海草床			
East Coast of Hainan	Hainan Province		Coral Reef Seagrass Bed			
西沙珊瑚礁	海南省	400	珊瑚礁			
Xisha Coral Reef	Hainan Province		Coral Reef			

注:*生物多样性指数：是生物种数和种类间个体数量分配均匀性的综合表现，用Shannon-Wiener多样性指数表征。

Note: Biodiversity index: refers to the comprehensive expression of the distributive homogeneity of the number of biological species and the number of individuals between varieties characterized by the Shannon-Wiener biodiversity index.

8-17 沿海地区风暴潮灾害情况
Survey of Storm Surges by Coastal Regions

受灾地区 Disaster Area	受灾人口（万人） Disaster-stricken Population (10 000 persons)	死亡人数*（人） Death Toll (person)	受灾面积 Disaster-stricken area	
			农田（千公顷） Disaster-Affected farmland (1 000 hm^2)	水产养殖（千公顷） Affected Area of Mariculture (1 000 hm^2)
合 计 Total	**1 380.34**	**0**	**354.66**	**140.16**
辽 宁 Liaoning		0	0	0
江 苏 Jiangsu		0	0	2.26
浙 江 Zhejiang	736.62	0	340.29	37.81
福 建 Fujian	38.07	0	1.11	54.85
山 东 Shandong		0	0	7.24
广 东 Guangdong	589.44	0	13.26	37.38
广 西 Guangxi	16.21	0	0	0.62

8-17 续表 continued

受灾地区 Disaster Area	海岸工程 （千米） Coastal Engineering (km)	房屋 House （间） (unit)	船只 （艘） Boats (unit)	直接经济损失 （亿元） Direct Economic Loss (100 million yuan)
合 计 Total	**284.27**	**6 308**	**13 717**	**153.96**
辽 宁 Liaoning	0.20	0	4	0.02
江 苏 Jiangsu	6.37	0	0	0.17
浙 江 Zhejiang	29.55	175	2 124	28.17
福 建 Fujian	125.02	1 101	6 066	45.06
山 东 Shandong	15.11	411	109	1.44
广 东 Guangdong	100.38	4 001	5 382	74.20
广 西 Guangxi	7.64	620	32	4.90

注：*包括失踪人数。

Notes:*Includes lost people.

8-18 沿海地区赤潮灾害情况
Survey of Red Tide by Coastal Regions

时间 Date	影响区域 Area	最大面积（平方千米） Disaster Area(km^2)
累　计 **In the Aggregate**		**3 277**
其中：Including:		
7月5日—7月8日 Jul.5-Jul.8	天津市临港经济区东部海域 Eastern Sea Area of the Port Economic Zone of Tianjin Municipality	154
7月16日—7月25日 Jul.16-Jul.25	天津市汉沽海域 Hangu Sea Area of Tianjin Municipality	100
5月25日—8月31日 May.25-Aug.31	河北省秦皇岛——绥中附近海域 Seawaters near Qinhuangdao-Suizhong of Hebei Province	1 450
5月30日—6月1日 May.30-Jun.1	江苏省连云港海州湾海域 Haizhou Bay Area of Lianyungang, Jiangsu Province	450
5月13日—5月29日 May.13-May.29	浙江省温州苍南海域 Cangnan Sea Area of Wenzhou, Zhejiang Province	450
5月18日—6月2日 May.18-Jun.2	浙江省台州玉环坎门海域 Kanmen Sea Area of Yuhuan, Taizhou, Zhejiang Province	120
5月20日—5月24日 May.20-May.24	浙江省宁波韭山列岛东南海域 Sea Area Southeast of the Jiushan Archipelago, Ningbo, Zhejiang Province	140
6月3日—6月9日 Jun.3-Jun.9	浙江省舟山东福山岛附近海域 Sea Area near Dongfushan Island, Zhoushan, Zhejiang Province	100
6月22日—6月24日 Jun.22-Jun.24	浙江省朱家尖岛东北部——中街山列岛西部海域 Western Seawaters Between the northeastern Zhujiajian Island-the Zhongjieshan Archipelago, Zhejiang Province	200
8月9日—8月13日 Aug.9-Aug.13	广东省湛江港湾附近海域 Sea Area near the Zhanjiang Harbour, Guangdong Province	113

主要统计指标解释

1. 工业废水排放量 指经过企业厂区所有排放口排到企业外部的工业废水量。包括生产废水、外排的直接冷却水、超标排放的矿井地下水和与工业废水混排的厂区生活污水,不包括外排的间接冷却水(清污不分流的间接冷却水应计算在内)。

2. 直接排入海的工业废水量 指经企业位于海边的排放口，直接排入海中的废水量。直接排入是指废水经过工厂的排污口直接排入海，而未经过城市下水道或其他中间体，也不受其他水体的影响。

3. 工业废水处理量 指报告期内各种水治理设施实际处理的工业废水量,包括处理后外排的和处理后回用的工业废水量，虽经处理但未达到国家或地方排放标准的废水量也应计算在内。计算时，如遇车间和厂排放口均有治理设施，并对同一废水分级处理时，不应重复计算工业废水处理量。

4. 一般工业固体废物处置量 指将固体废物焚烧或者最终置于符合环境保护规定要求的场所并不再回取的工业固体废物量(包括当年处置往年的工业固体废物累计贮存量)。

5. 一般工业固体废物倾倒丢弃量 指将所产生的固体废物倾倒丢弃到固体废物污染防治设施、场所以外的量。不包括矿山开采的剥离废石和掘进废石(煤矸石和呈酸性或碱性的废石除外)。

6. 当年安排施工项目数 指报告期内由国家、部门、地方或企业单位安排开工的，并以治理废水、废气、固体废物、噪声和其他（如电磁波、恶臭等）环境污染的环境治理工程的总数。不包括“三同时”项目。

7. 当年竣工项目数 指报告期内竣工投入运行的治理废水、废气、固体废物、噪声及其他污染的环境工程项目的总数。

Explanatory Notes on Main Statistical Indicators

1. Volume of Industrial Waste Water Discharged refers to the quantity of industrial waste water discharged externally through all the outlets in the factory area of the enterprise, including the waste water from production, externally discharged direct cooling water, mine-shaft groundwater discharged exceeding the set standard and the domestic sewage of the factory area discharged together with the industrial waste water, but not including the indirect cooling water discharge externally.

2. Volume of Industrial Waste Water discharged Directly to Sea refers to the quantity of waste water directly discharged into the sea through the outlets of the enterprise by the sea. Direct discharge means the direct discharge into the sea of waste water through the outlets of the factory, which is not discharged via the urban sewers or other intermediates and is not affected by other water bodies.

3. Volume of Industrial Waste Water Treated refers to the industrial waste water volume actually treated by various water treatment facilities in the period covered by the report, including the quantity of the industrial waste water discharged and reused after treatment. The amount of the waste water which is not up to the state or local standard of discharge upon treatment should be included. If the workshops and outlets of the factory are provided with treatment facilities and carry out graded treatment of the same waste water, the processed volume of industrial waste water cannot be calculated repeatedly.

4. Volume of Common Industrial Solid Wastes Discharged refers to the amount of solid wastes discharged out of the facilities and sites for the solid wastes pollution prevention and control, not including the stripped and tunneled waste ores in the excavation of mines (other than gangues and acid or alkaline waste ores).

5. Volume of Common Industrial Solid Wastes Treated refers to the volume of industrial solid wastes which are to be burned or finally placed at the sites in keeping with the requirement of environmental protection and will not be recovered (including the accumulated amount of storage in former years of industrial solid radioactive matter disposed of in the year).

6. Arranged for Construction in the Year refers to the total number of environmental pollution control projects for controlling waste water, waste gas, solid wastes, noise and other environmental pollutions (such as electromagnetic wave, offensive odor) started by the state, governmental departments, local governments or enterprises in the period covered by the report mainly for the purpose of pollution control and multipurpose utilization of "three wastes".

7. Number of Pollution Treatment Projects Completed in the Current Year refers to the total number of environmental engineering projects for controlling waste water, waste gas, solid wastes, noise and other pollutions which are completed and put into operation in the period covered by the report.

9 海洋行政管理及公益服务

Marine Administration and Public-Good Service

9-1 海域使用管理情况
Sea Area Use Management

地 区 Region	颁发海域使用权证书（本） Certificates of Right of Sea Area Use Issued (volume)	确权海域面积（公顷） Area of Waters with Established Rights (hm^2)	征收海域使用金（万元） Charge for Sea Area Utilization (10 000 yuan)
全国总计 National Total	**3 937**	**354 979**	**1 089 241**
天 津 Tianjin	45	1 502	105 237
河 北 Hebei	80	1 701	47 437
辽 宁 Liaoning	861	144 861	143 239
上 海 Shanghai	4	79	1 453
江 苏 Jiangsu	294	51 205	80 428
浙 江 Zhejiang	319	7 262	281 278
福 建 Fujian	313	7 697	55 732
山 东 Shandong	1 141	126 680	146 269
广 东 Guangdong	244	6 150	98 384
广 西 Guangxi	500	4 684	39 599
海 南 Hainan	131	2 915	79 633
其 他* other	5	243	10 552

注：*为沿海省（自治区、直辖市）管理海域以外（指渤海中部海域）。

Note: * Sea areas outside the control of coastal provinces of autonomous regions and municipalities directly under the Central Government. (Referring to the mid-Bohai Sea area)

9-2 海洋倾废管理情况
Management on the Ocean Dumping of Wastes

海 区 Sea Area	签发疏浚物海洋倾倒许可证（份） Permit Issued for Ocean Dumping of Dredged Materials (unit)	新选划倾倒区数（个） Number of Newly Designated Dumping Zones (unit)
合 计 Total	**116**	**7**
渤黄海 Bohai and Huanghai Sea	38	5
东 海 East China Sea	28	2
南 海 South China Sea	50	

9-3 海洋执法检查情况
Marine Law Enforcement Inspection

执法领域 Law Enforcement Area	检查项目（个）Number of Items Subject to Inspection (number)	检查次数（次）Number of Times of Inspection (number)	发现违法行为（起）Illegal Acts Found (case)	作出行政处罚（件）Inflict Administrative Punishments (case)
总　计 Total	**53 412**	**166 588**	**2 228**	**1 294**
海域使用 Sea Area Use	26 718	94 556	1 399	609
涉外海洋科研 Foreign-related Marine Scientific Research	45	51	0	0
海底电缆保护 Submarine Cable Protection	139	882	5	4
海洋工程建设项目 Marine Engineering Construction Projects	7 669	32 907	290	254
海洋倾废 Oceanic Dumping of Wastes	1 524	6 623	173	160
海洋生态保护 Marine Ecological Conservation	659	4 832	311	244
海岛保护 Island Protection	16 658	26 737	50	23

9-4 沿海地区海滨观测台站分布概况
Distribution of Coastal Observation Stations by Coastal Regions

单位：个 (unit)

地　区 Region	合　计 Total	海洋站 Marine Station	验潮站[①] Tide Station	气象台站 Meteorological Station	地震台站 Seismic Station	雷达站 Radar Station
合　计 Total	**1 200**	**108**	**251**	**416**	**282**	**143**
天　津 Tianjin	**26**	1	0	14	7	4
河　北 Hebei	**34**	4	0	3	21	6
辽　宁 Liaoning	**103**	7	3	53	29	11
上　海 Shanghai	**93**	7	62		10	14
江　苏 Jiangsu	**90**	8	26	34	18	4
浙　江 Zhejiang	**139**	20	37	42	21	19
福　建 Fujian	**213**	15	11	119	47	21
山　东 Shandong	**153**	16	22	53	39	23
广　东 Guangdong	**212**	16	82	32	56	26
广　西 Guangxi	**43**	5	5	19	8	6
海　南 Hainan	**94**	9	3	47	26	9

注：①潮流量观测站43处，潮水位观测站208处。

Notes: ①There are 43 tidal current observation stations and 208 tidal level observation stations.

9-5 海洋预报服务概况
Marine Forecast Service

单位：次 (time)

预报项目 Item	数值预报 Numerical Forecast			
	预报服务次数 Frequency	发布次数 Frequency of Release		
		广播电视 Radio and TV	互联网 Internet	纸 质 Paper Media
合 计 Total	**76 082**	**1 825**	**73 962**	**295**
海 浪 Sea Wave	12 317	1 825	10 370	122
海 温 Sea Surface Temperature	1 582		1 460	122
潮 汐 Tide	4 594		4 594	
海 流 Sea Current	6 205		6 205	
海平面 Sea Level				
盐 度 Salinity	1 095		1 095	
赤 潮 Red Tide				
滨海旅游 Coastal Tourism				
海 冰 Sea Ice	140		140	
绿 潮 Green Seaweed	1 470		1 470	
溢 油 Oil Spill	7		5	2
厄尔尼诺 EL Niño				
专 项 Special Item	48 075		48 059	16
其 他 Others	597		564	33

9-5 续表 continued

预报项目 Item	统计预报 Statistical Forecast			
	预报服务次数 Frequency	发布次数 Frequency of Release		
		广播电视 Radio and TV	互联网 Internet	纸 质 Paper Media
合 计 Total	**176 666**	**54 091**	**69 150**	**53 425**
海 浪 Sea Wave	50 987	19 411	23 255	8 321
海 温 Sea Surface Temperature	41 321	15 500	20 064	5 757
潮 汐 Tide	36 889	15 029	15 514	6 346
海 流 Sea Current				
海平面 Sea Level				
盐 度 Salinity				
赤 潮 Red Tide	210	6	90	114
滨海旅游 Coastal Tourism	9 006	3 630	4 178	1 198
海 冰 Sea Ice	2 283	120	175	1 988
绿 潮 Green Seaweed	60			60
溢 油 Oil Spill	30			30
厄尔尼诺 EL Niño				
专 项 Special Item	35 119	395	5 490	29 234
其 他 Others	761		384	377

注：本表只包含国家海洋局资料。

Note: This table only contains the data from the State Oceanic Administration.

9-6 海洋环境观测情况
Condition of Marine Environmental Observation

项目 Item	台站观测 Station Observation	断面观测 Sectional Observation	浮标观测 Buoy Observation	船舶测报 Ship Measuring and Reporting	其他观测 Other Observation
测站（点）数（个） Number of Stations (unit)	108	120	67	95	132
实际获得数据量（个） Quantity of Data Actually Obtained (unit)	272 599 939	10 109	4 022 687	16 475 852	229 684 377

9-7 海洋调查概况
Marine Survey Statistics

调查名称 Name	站点数（个） Number of Stations (unit)	船舶数（艘） Number of Ships (unit)	项目数（个） Number of Items (unit)	实际获得数据（个） Quantity of Data Actually Obtained (unit)	发布通（公、简）报量（期） Quantity of Circulars (Bulletin, Brief Reports) (unit)
合　计 Total	**3 203**	**352**	**569**	**2 329 579**	**94**
大洋调查 Oceanic Survey	375	7	30	53 447	65
极地调查 Polar Survey	273	3	28	8 943	
专项调查 Special Survey	1 689	45	100	1 908 138	6
其他调查 Other Surveys	866	297	411	359 051	23

9-8 涉外海洋科学研究审批情况
Examination and Approval of Foreign-Related Marine Scientific Research

单位：份 (unit)

审批单位 Examing and Approving Units	审批研究活动申请 Application for Examing and Approving Research Activities	船只作业计划审批 Examination and Approval of the Ship Operating Plan
国家海洋局 State Oceanic Administration , People's Republic of China	4	4

9-9 海洋档案及利用情况
Marine File and Its Use

指　标 Item	指标值 Data
室（馆）存档案 Files Deposited in the Archives	
纸介质档案（卷、册） Paper Media Files (reel,volume)	135 637
电子档案（GB） Electronic Media Files (GB)	18 078
本年接收档案 Files Received this Year	
纸质档案（卷、册） Paper Media Files (reel,volume)	23 952
电子档案（GB） Electronic Media Files (GB)	4 089
本年利用档案 Files Utilized this Year	
纸质档案（件） Paper Media Files (piece)	25 167
电子档案（GB） Electronic Media Files (GB)	4 085

9-10 卫星遥感接收应用情况
Remote-Sensing Receiving and Utilization

指 标 Item	指标值 Data
卫星接收次数（轨） Number of Satellite Receptions (orbit / track)	54 807
全年接收时间（天） Whole Year Receiving Time (day)	6 557
全年实际接收存档数据量（GB） Amount of Data Actually Received and Placed on File in the Whole Year (GB)	25 714.28
累计存档数据量（GB） Total Amount of Data Placed on File (GB)	42 640.61
卫星数据分发 Satellite Data Distribution	
类别用户（个） Classified Users (unit)	
分发数据量（GB） Amount of Data Distributed (GB)	23 390.80

9-11 海洋标准化监督管理情况
Supervision and Management of Marine Standardization

单位：项 (item)

指 标 Item	指标值 Data
标准立项审查 Examination of the Standards for Authorization	
国家标准 National Standards	21
行业标准 Professional Standards	72
标准审查 Examination of Standards	
国家标准 National Standards	52
行业标准 Professional Standards	112
标准出版 Standards Publication	
国家标准 National Standards	1
行业标准 Professional Standards	41
标准实施监督检查（次） Supervision and Examination of Standards Implementation (time)	16

主要统计指标解释

1. 海域使用检查 针对不同类型的用海行为进行的监督检查。

2. 涉外海洋科研项目检查 主要针对国际组织、外国组织和个人为和平目的，单独或者与中华人民共和国的组织合作，使用船舶或者其他运载工具、设施，在中华人民共和国内海、领海以及中华人民共和国管辖的其他海域内进行的对海洋环境和海洋资源等的调查研究活动进行监督检查。

3. 海底电缆管道检查 主要是针对铺设海底电缆管道路由调查、铺设施工和维修改造等的监督检查。

4. 海洋工程建设项目环境保护检查 主要是针对防治海洋工程建设项目对海洋环境的污染损害的监督检查。

5. 海洋倾废检查 主要是针对防治倾倒废弃物对海洋环境的污染损害的监督检查。

6. 海洋生态保护检查 主要是针对红树林、珊瑚礁、滨海湿地、海岛、海湾、入海河口、重要渔业水域等具有典型性、代表性的海洋生态系统，珍稀、濒危海洋生物的天然集中分布区，具有重要经济价值的海洋生物生存区域及有重大科学文化价值的海洋自然历史遗迹和自然景观等海洋自然保护区以及其他需要予以特殊保护的区域的监督检查。

7. 断面监测 按照国家海洋局“断面监测方案”，每年定期利用船舶在沿海设定的断面上进行海洋水文、气象、生物、化学等项目的监测活动。

8. 浮标监测 在海上固定站位获取长期、连续海洋环境观测资料的海上锚定资料浮标。

9. 船舶监测 利用在固定航线上走航的商船或渔船，每日定时所在地的海洋环境状况（主要是水文气象要素），并将监测数据实时发送给有关单位，供海洋环境预报使用。

10. 大洋调查 以大洋科考、研究为目的的远洋调查。

11. 专项调查 为完成国家专项任务进行的海洋调查。

12. 全年接收时间 指全年整个应用系统接收的各类遥感卫星数据时，接收设备工作时间。

13. 全年实际接收存档数据量 指全年整个应用系统接收的各类遥感卫星数据原始数据量。

14. 累计存档数据量 指全年整个应用系统制作的各类遥感产品数据量。

15. 分发数据量 指应用系统提供数据产品用于开展各类应用的数据量。

16. 卫星接收次数 地面观测系统接收卫星数据的数量。

17. 国家标准 针对海洋领域内需要在全国范围内统一的有关技术要求所制定的国家标准。海洋国家标准由国家标准化主管部门统一批准、编号和发布。

18. 行业标准 对没有海洋国家标准而又需要在海洋领域内统一的技术要求所制定的标准。海洋行业标准由国家海洋局统一批准、编号和发布。

Explanatory Notes on Main Statistical Indicators

1. Sea Area Use Supervision and inspection refers to the various types of sea area use conducts.

2. Inspection of Foreign-Related Marine Scientific Research Projects refers to the supervision and inspection of the activities of surveying marine environment and resources conducted by international organizations, foreign organizations and individuals for peaceful purposes, alone or in cooperation with PRC organizations, by using ships or other means of delivery as facilities in the internal seas and territorial waters of the People's Republic of China as well as in the other water under the jurisdiction of the People's Republic of China.

3. Inspection of Submarine Cables and Pipelines is mainly aimed at the survey of the submarine cables and pipelines laying route and supervision and inspection of the laying construction, repair and transformation.

4. Environmental Protection Inspection for the Marine Engineering Construction Projects is mainly directed against preventing and controlling the pollution damage of the marine engineering construction project to the marine environment.

5. Inspection of Oceanic Dumping of Wastes mainly refers to the supervision and inspection aimed at preventing and controlling the pollution damage of wastes dumping to the marine environment.

6. Inspection of Marine Ecological Protection mainly refers to the supervision and inspection of the typical and representative marine ecosystems such as mangroves, coral reef, coastal wetland, sea island, bay, estuaries open to the sea, important fishery waters, etc, the natural centralized distribution zones of rare and endangered marine life, the living areas for the marine life with significant economic values and the marine nature reserves such as the marine natural and historical remains and natural landscapes with important scientific and cultural values as well as other areas needing to be specially protected.

7. Sectional Monitoring According to the *"Sectional Monitoring Plan"* of the State Oceanic Administration, monitoring activities concerning such items as marine hydrology, meteorology, biology and chemistry are carried out regularly every year on the sections set in the coastal area.

8. Buoy Monitoring refers to the monitoring carried out by the offshore mooring data buoys which acquire long-term, continuous marine environmental observations at the fixed stations at sea.

9. Ship Monitoring Monitoring the marine environmental condition in the sea area of fixed times every day by using the merchant or fishing vessels cruising on the fixed navigation line (mainly the hydro meteorological elements) and transmitting the monitored data in real time to the related units for use in the marine environmental forecast.

10. Oceanic Survey refers to the oceanic surveys aimed at the oceanic scientific investigations and research.

11. Special-Subject Survey refers to the oceanic investigation for the purpose of fulfilling the state's special tasks

12. Whole Year Receiving Time refers to the operating time of receiving equipment in the whole year when the whole application system receives all types of remote sensing satellite data.

13. Amount of Data Actually Received and Placed on File Throughout the year refers to the raw data of remote-sensing satellite data of various kinds received by the whole application system throughout the year.

14. Total Amount of Date Placed on File refers to the date amount of various remote-sensing products made by the whole application systems throughout the year.

15. Amount of Data Distributed refers to the amount of data products provided by the application system for various uses.

16. Number of Satellite Receptions refers to the amount of data received by the ground observation system.

17. National Standards refers to the standards formulated in view of the relevant technical requirements in the marine field that need to be unified throughout the country. The Marine National standards are approved, numbered and issued uniformly by the state department responsible for standardization.

18. Professional Standards refers to the standards formulated for the technical requirements which have no national standards but need to be unified in the marine field. The Marine Professional Standards are approved, numbered and issued by the State Oceanic Administration.

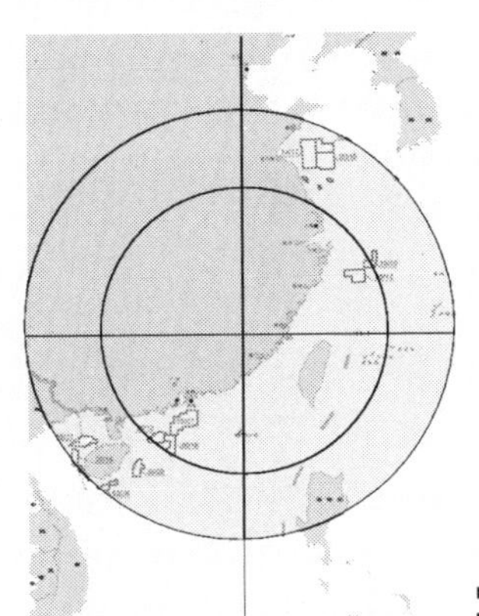

10

全国及沿海社会经济

National and Coastal Socioeconomy

10-1 国内生产总值
Gross Domestic Product

单位：亿元 (100 million yuan)

年 份 Year	国内生产总值 Gross Domestic Product	第一产业 Primary Industry	第二产业 Secondary Industry	第三产业 Tertiary Industry
2001	109 655.2	15 781.3	49 512.3	44 361.6
2002	120 332.7	16 537.0	53 896.8	49 898.9
2003	135 822.8	17 381.7	62 436.3	56 004.7
2004	159 878.3	21 412.7	73 904.3	64 561.3
2005	184 937.4	22 420.0	87 598.1	74 919.3
2006	216 314.4	24 040.0	103 719.5	88 554.9
2007	265 810.3	28 627.0	125 831.4	111 351.9
2008	314 045.4	33 702.0	149 003.4	131 340.0
2009	340 902.8	35 226.0	157 638.8	148 038.0
2010	401 512.8	40 533.6	187 383.2	173 596.0
2011	473 104.0	47 486.2	220 412.8	205 205.0
2012	519 470.1	52 373.6	235 162.0	231 934.5
2013	568 845.2	56 957.0	249 684.4	262 203.8

注：2013年为初步核算数据。
Note: Data of 2013 were preliminary estimation.

10-2 国内生产总值增长速度
Growth Rate of Gross Domestic Product

单位：% (%)

年 份 Year	国内生产总值 Gross Domestic Product	第一产业 Primary Industry	第二产业 Secondary Industry	第三产业 Tertiary Industry
2001	8.3	2.8	8.4	10.3
2002	9.1	2.9	9.8	10.4
2003	10.0	2.5	12.7	9.5
2004	10.1	6.3	11.1	10.1
2005	11.3	5.2	12.1	12.2
2006	12.7	5.0	13.4	14.1
2007	14.2	3.7	15.1	16.0
2008	9.6	5.4	9.9	10.4
2009	9.2	4.2	9.9	9.6
2010	10.4	4.3	12.3	9.8
2011	9.3	4.3	10.3	9.4
2012	7.7	4.5	7.9	8.1
2013	7.7	4.0	7.8	8.3

注：本表按可比价格计算(上年为基期)。
Note:This table is calculated at the comparable price (with the previous year as the base period).

10-3 沿海地区生产总值
Gross Regional Product of Coastal Regions

单位：亿元 (100 million yuan)

地 区 Region	地区生产总值 Gross Regional Product	第一产业 Primary Industry	第二产业 Secondary Industry	第三产业 Tertiary Industry
合 计 Total	**344 213.98**	**24 396.97**	**167 776.28**	**152 040.73**
天 津 Tianjin	14 370.16	188.45	7 276.68	6 905.03
河 北 Hebei	28 301.41	3 500.42	14 762.10	10 038.89
辽 宁 Liaoning	27 077.65	2 321.63	14 269.46	10 486.56
上 海 Shanghai	21 602.12	129.28	8 027.77	13 445.07
江 苏 Jiangsu	59 161.75	3 646.08	29 094.03	26 421.64
浙 江 Zhejiang	37 568.49	1 784.62	18 446.65	17 337.22
福 建 Fujian	21 759.64	1 936.31	11 315.30	8 508.03
山 东 Shandong	54 684.33	4 742.63	27 422.47	22 519.23
广 东 Guangdong	62 163.97	3 047.51	29 427.49	29 688.97
广 西 Guangxi	14 378.00	2 343.57	6 863.04	5 171.39
海 南 Hainan	3 146.46	756.47	871.29	1 518.70

10-4 沿海地区生产总值增长速度
Growth Rate of Gross Regional Product of Coastal Regions

单位：% (%)

地 区 Region	2002	2003	2004	2005	2006	2007	2008	2009	2010	2011	2012	2013
天 津 Tianjin	12.7	14.8	15.8	14.9	14.7	15.5	16.5	16.5	17.4	16.4	13.8	12.5
河 北 Hebei	9.6	11.6	12.9	13.4	13.4	12.8	10.1	10.0	12.2	11.3	9.6	8.2
辽 宁 Liaoning	10.2	11.5	12.8	12.7	14.2	15.0	13.4	13.1	14.2	12.2	9.5	8.7
上 海 Shanghai	11.3	12.3	14.2	11.4	12.7	15.2	9.7	8.2	10.3	8.2	7.5	7.7
江 苏 Jiangsu	11.7	13.6	14.8	14.5	14.9	14.9	12.7	12.4	12.7	11.0	10.1	9.6
浙 江 Zhejiang	12.6	14.7	14.5	12.8	13.9	14.7	10.1	8.9	11.9	9.0	8.0	8.2
福 建 Fujian	10.2	11.5	11.8	11.6	14.8	15.2	13.0	12.3	13.9	12.3	11.4	11.0
山 东 Shandong	11.7	13.4	15.4	15.0	14.7	14.2	12.0	12.2	12.3	10.9	9.8	9.6
广 东 Guangdong	12.4	14.8	14.8	14.1	14.8	14.9	10.4	9.7	12.4	10.0	8.2	8.5
广 西 Guangxi	10.6	10.2	11.8	13.1	13.6	15.1	12.8	13.9	14.2	12.3	11.3	10.2
海 南 Hainan	9.6	10.6	10.7	10.5	13.2	15.8	10.3	11.7	16.0	12.0	9.1	9.9

注：本表按可比价格计算(上年为基期)。

Note: This table is calculated at the comparable price (with the previous year as the base period).

10-5 沿海城市生产总值（2012年）
Gross Regional Product of Coastal Cities, 2012

单位：亿元 (100 million yuan)

沿海城市 Coastal City		地区生产总值 Gross Regional Product	第一产业 Primary Industry	第二产业 Secondary Industry	第三产业 Tertiary Industry
合 计	**Total**	**188 770.11**	**9 886.11**	**90 937.03**	**87 947.01**
天 津	**Tianjin**	**14 370.16**	**188.45**	**7 276.68**	**6 905.03**
河 北	**Hebei**	**9 813.43**	**1 000.15**	**5 400.55**	**3 412.73**
唐 山	Tangshan	5 861.64	528.56	3 473.79	1 859.29
秦皇岛	Qinhuangdao	1 139.37	152.41	447.68	539.27
沧 州	Cangzhou	2 812.42	319.18	1 479.08	1 014.17
辽 宁	**Liaoning**	**12 606.38**	**1 089.57**	**6 684.99**	**4 831.83**
大 连	Dalian	7 002.83	451.37	3 634.80	2 916.66
丹 东	Dandong	1 015.37	140.01	508.53	366.84
锦 州	Jinzhou	1 242.71	190.29	616.23	436.19
营 口	Yingkou	1 381.18	103.56	738.99	538.63
盘 锦	Panjin	1 244.96	108.43	843.55	292.97
葫芦岛	Huludao	719.33	95.91	342.89	280.54
上 海	**Shanghai**	**21 602.12**	**129.28**	**8 027.77**	**13 445.07**
江 苏	**Jiangsu**	**9 282.09**	**1 007.62**	**4 623.12**	**3 651.35**
南 通	Nantong	4 558.67	319.09	2 414.11	1 825.47
连云港	Lianyungang	1 603.42	232.40	736.14	634.88
盐 城	Yancheng	3 120.00	456.13	1 472.87	1 191.00
浙 江	**Zhejiang**	**28 362.44**	**1 258.01**	**14 310.36**	**12 794.07**
杭 州	Hangzhou	7 802.01	255.11	3 572.63	3 974.27
宁 波	Ningbo	6 582.21	268.52	3 516.84	2 796.85
温 州	Wenzhou	3 669.18	114.22	1 852.99	1 701.98
嘉 兴	Jiaxing	2 890.57	151.39	1 603.08	1 136.10
绍 兴	Shaoxing	3 654.03	184.80	1 962.41	1 506.82
舟 山	Zhoushan	853.18	83.06	382.94	387.18
台 州	Taizhou	2 911.26	200.91	1 419.47	1 290.87
福 建	**Fujian**	**16 017.16**	**1 182.64**	**8 322.74**	**6 511.79**
福 州	Fuzhou	4 210.93	367.73	1 905.50	1 937.70
厦 门	Xiamen	2 815.17	25.30	1 363.85	1 426.02
莆 田	Putian	1 200.38	107.24	689.65	403.49
泉 州	Quanzhou	4 702.70	160.57	2 890.41	1 651.72
漳 州	Zhangzhou	2 012.92	320.45	961.10	731.37
宁 德	Ningde	1 075.06	201.35	512.23	361.49

10-5 续表 continued

沿海城市 Coastal City		地区生产总值 Gross Regional Product	第一产业 Primary Industry	第二产业 Secondary Industry	第三产业 Tertiary Industry
山 东	**Shandong**	**25 274.74**	**1 683.84**	**13 698.48**	**9 892.42**
青 岛	Qingdao	7 302.11	324.41	3 402.23	3 575.47
东 营	Dongying	3 000.66	104.34	2 126.02	770.30
烟 台	Yantai	5 281.38	377.31	2 985.09	1 918.98
潍 坊	Weifang	4 012.43	390.52	2 166.17	1 455.74
威 海	Weihai	2 337.86	180.11	1 249.30	908.45
日 照	Rizhao	1 352.57	117.64	724.06	510.87
滨 州	Binzhou	1 987.73	189.51	1 045.61	752.61
广 东	**Guangdong**	**48 526.47**	**1 888.56**	**21 496.03**	**25 141.90**
广 州	Guangzhou	13 551.21	213.76	4 720.65	8 616.79
深 圳	Shenzhen	12 950.06	6.30	5 737.64	7 206.12
珠 海	Zhuhai	1 503.76	39.02	776.36	688.38
汕 头	Shantou	1 425.01	80.44	740.20	604.37
江 门	Jiangmen	1 880.39	149.51	960.82	770.06
湛 江	Zhanjiang	1 860.22	384.97	723.10	752.15
茂 名	Maoming	1 936.18	344.12	789.55	802.51
惠 州	Huizhou	2 367.55	124.56	1 377.23	865.76
汕 尾	Shanwei	610.41	99.28	284.18	226.95
阳 江	Yangjiang	887.03	176.02	408.93	302.08
东 莞	Dongguan	5 010.17	18.76	2 375.64	2 615.78
中 山	Zhongshan	2 441.04	62.17	1 353.64	1 025.24
潮 州	Chaozhou	706.65	49.44	388.74	268.48
揭 阳	Jieyang	1 396.79	140.21	859.35	397.23
广 西	**Guangxi**	**1 765.40**	**355.34**	**826.46**	**583.60**
北 海	Beihai	630.09	127.37	303.75	198.97
防城港	Fangchenggang	443.99	61.16	233.56	149.28
钦 州	Qinzhou	691.32	166.81	289.15	235.35
海 南	**Hainan**	**1 149.72**	**102.65**	**269.85**	**777.22**
海 口	Haikou	818.76	55.92	201.67	561.17
三 亚	Sanya	330.96	46.73	68.18	216.05

注：本表各省数据为合计数。

Note: The data for the provinces are the totals.

10-6 沿海县主要统计指标（2012年）
Main Statistical Indicators of Coastal Counties, 2012

单位：万元 (10 000 yuan)

沿海县 Coastal County		第一产业增加值 Value-added of Primary Industry	第二产业增加值 Value-added of Secondary Industry	地方财政一般预算收入 General Budget Revene of Local Governments	地方财政一般预算支出 General Budget Expenditure of Local Governments
合　计	**Total**	**50 650 179**	**207 016 349**	**26 227 677**	**40 735 290**
河　北	**Hebei**	**2 849 599**	**4 577 105**	**410 615**	**1 065 897**
丰　南	Fengnan				
滦　南	Luannan	754 496	1 143 063	89 880	201 926
乐　亭	Leting	713 622	1 037 805	81 295	227 126
唐　海	Tanghai				
昌　黎	Changli	636 621	691 922	69 157	181 483
抚　宁	Funing	424 287	558 864	54 034	151 429
黄　骅	Huanghua	257 919	1 010 481	94 534	220 858
海　兴	Haixing	62 654	134 970	21 715	83 075
辽　宁	**Liaoning**	**6 711 184**	**23 794 685**	**3 147 288**	**4 325 306**
长　海	Changhai	504 796	78 169	42 009	67 794
瓦房店	Wafangdian	940 154	6 262 866	641 174	762 440
普兰店	Pulandian	967 893	4 324 593	419 964	486 885
庄　河	Zhuanghe	1 149 566	4 473 355	482 357	581 231
东　港	Donggang	691 485	2 275 791	300 006	411 512
凌　海	Linghai	498 460	1 462 651	215 050	329 222
盖　州	Gaizhou	352 928	1 061 060	210 617	304 289
大　洼	Dawa	606 642	1 795 253	356 980	489 924
盘　山	Panshan	441 749	860 369	207 483	315 909
绥　中	Suizhong	373 729	600 642	154 107	325 800
兴　城	Xingcheng	183 782	599 936	117 541	250 300

10-6 续表1 continued

沿海县 Coastal County		第一产业增加值 Value-added of Primary Industry	第二产业增加值 Value-added of Secondary Industry	地方财政一般预算收入 General Budget Revene of Local Governments	地方财政一般预算支出 General Budget Expenditure of Local Governments
江　苏	**Jiangsu**	**6 999 907**	**24 167 928**	**4 308 020**	**6 558 208**
海　安	Hai'an	476 897	2 463 012	375 279	552 958
如　东	Rudong	565 628	2 430 255	321 681	544 423
启　东	Qidong	614 819	3 060 885	522 149	595 882
海　门	Haimen	464 998	3 775 976	516 092	559 553
赣　榆	Ganyu	504 600	1 660 700	292 104	536 889
东　海	Donghai	510 200	1 273 700	274 607	480 020
灌　云	Guanyun	503 400	1 022 300	258 598	410 191
灌　南	Guannan	393 900	1 052 800	255 941	418 773
响　水	Xiangshui	355 840	898 500	197 301	358 616
滨　海	Binhai	497 010	1 178 900	249 441	480 926
射　阳	Sheyang	695 917	1 293 800	208 036	438 319
东　台	Dongtai	783 298	2 318 900	436 725	627 508
大　丰	Dafeng	633 400	1 738 200	400 066	554 150
浙　江	**Zhejiang**	**6 041 431**	**49 877 213**	**6 829 448**	**8 396 110**
象　山	Xiangshan	537 669	1 583 874	272 793	458 059
宁　海	Ninghai	364 069	1 948 447	298 439	417 143
余　姚	Yuyao	434 637	4 223 347	616 855	678 191
慈　溪	Cixi	465 351	5 600 405	813 852	866 031
奉　化	Fenghua	276 749	1 303 059	243 728	389 930
洞　头	Dongtou	38 102	186 776	33 926	131 756
平　阳	Pingyang	134 083	1 285 873	171 174	275 092
苍　南	Cangnan	260 866	1 540 191	193 273	380 704
瑞　安	Rui'an	188 009	2 763 596	406 484	466 178
乐　清	Yueqing	196 886	3 489 699	457 238	464 758
海　盐	Haiyan	210 114	1 770 746	223 273	234 559
海　宁	Haining	252 980	3 396 264	442 890	433 163
平　湖	Pinghu	178 672	2 623 718	380 180	374 403

10-6 续表2 continued

沿海县 Coastal County		第一产业增加值 Value-added of Primary Industry	第二产业增加值 Value-added of Secondary Industry	地方财政一般预算收入 General Budget Revene of Local Governments	地方财政一般预算支出 General Budget Expenditure of Local Governments
绍　兴	Shaoxing	356 485	5 893 289	704 424	626 082
上　虞	Shangyu	392 446	3 214 694	391 888	399 217
岱　山	Daishan	235 238	933 546	97 712	251 024
嵊　泗	Shengsi	160 774	102 758	48 629	142 702
玉　环	Yuhuan	253 094	2 249 488	251 128	296 662
三　门	Sanmen	210 039	548 143	106 818	206 448
温　岭	Wenling	535 290	3 244 831	388 678	486 941
临　海	Linhai	359 878	1 974 469	286 066	417 067
福　建	**Fujian**	**6 789 176**	**36 410 194**	**3 865 670**	**5 780 999**
连　江	Lianjiang	936 921	1 040 100	207 185	291 281
罗　源	Luoyuan	251 364	999 800	94 816	142 080
平　潭	Pingtan	331 656	461 400	104 152	461 233
福　清	Fuqing	820 180	3 062 800	390 257	480 520
长　乐	Changle	376 367	2 876 000	234 240	297 714
仙　游	Xianyou	258 376	961 356	112 861	256 003
惠　安	Hui'an	267 737	3 488 950	245 194	342 942
石　狮	Shishi	182 372	2 940 677	306 932	373 885
晋　江	Jinjiang	182 046	8 150 504	812 123	815 191
南　安	Nan'an	224 912	4 183 554	343 789	463 130
云　霄	Yunxiao	227 613	458 932	38 385	122 302
漳　浦	Zhangpu	535 046	810 567	140 901	291 075
诏　安	Zhao'an	340 702	544 920	44 072	138 908
东　山	Dongshan	270 333	515 408	83 008	176 292
龙　海	Longhai	539 516	2 697 244	352 123	463 450
霞　浦	Xiapu	395 098	419 261	59 717	169 338
福　安	Fu'an	344 014	1 604 327	158 350	261 282
福　鼎	Fuding	304 923	1 194 394	137 565	234 373

10-6 续表3 continued

沿海县 Coastal County			第一产业增加值 Value-added of Primary Industry	第二产业增加值 Value-added of Secondary Industry	地方财政一般预算收入 General Budget Revene of Local Governments	地方财政一般预算支出 General Budget Expenditure of Local Governments
山	**东**	**Shandong**	**7 800 914**	**49 192 223**	**5 131 665**	**6 986 479**
胶	州	Jiaozhou	476 822	4 188 500	450 617	513 198
即	墨	Jimo	568 004	4 176 500	506 206	614 223
胶	南	Jiaonan				
垦	利	Kenli	166 021	1 919 647	145 696	222 758
利	津	Lijin	251 612	1 116 473	74 638	182 805
广	饶	Guangrao	368 230	4 290 114	272 059	377 623
长	岛	Changdao	364 797	54 000	8 058	47 414
龙	口	Longkou	292 007	5 248 181	585 686	632 145
莱	阳	Laiyang	370 217	1 496 087	80 337	180 708
莱	州	Laizhou	586 033	3 237 031	360 007	454 732
蓬	莱	Penglai	250 561	2 390 454	236 098	301 643
招	远	Zhaoyuan	314 828	3 191 565	338 100	413 150
海	阳	Haiyang	511 844	1 270 853	175 116	261 078
寿	光	Shouguang	784 958	3 126 767	550 006	637 474
昌	邑	Changyi	312 913	1 757 755	186 618	251 506
文	登	Wendeng	499 987	3 744 270	325 084	472 679
荣	成	Rongcheng	702 221	4 200 341	426 588	744 236
乳	山	Rushan	312 087	2 011 965	200 018	296 650
无	棣	Wudi	344 253	1 201 974	126 525	212 380
沾	化	Zhanhua	323 519	569 746	84 208	170 077
广	**东**	**Guangdong**	**7 630 987**	**14 849 391**	**1 333 773**	**3 934 039**
南	澳	Nan'ao	34 799	51 081	12 980	95 349
台	山	Taishan	473 990	1 834 412	176 125	302 317
恩	平	Enping	174 357	470 784	70 528	166 956
遂	溪	Suixi	234 026	649 700	52 230	225 853
徐	闻	Xuwen	574 481	161 798	34 792	163 750

10-6 续表4 continued

沿海县 Coastal County		第一产业增加值 Value-added of Primary Industry	第二产业增加值 Value-added of Secondary Industry	地方财政一般预算收入 General Budget Revene of Local Governments	地方财政一般预算支出 General Budget Expenditure of Local Governments
廉　江	Lianjiang	793 850	971 686	67 338	289 688
雷　州	Leizhou	836 687	290 191	53 210	272 855
吴　川	Wuchuan	793 850	641 205	46 188	210 251
电　白	Dianbai	717 652	1 169 507	101 450	305 069
惠　东	Huidong	344 415	1 599 511	172 029	343 287
海　丰	Haifeng	310 951	893 816	113 459	197 107
陆　丰	Lufeng	428 477	786 590	121 358	317 116
阳　西	Yangxi	447 509	432 035	39 106	150 200
阳　东	Yangdong	359 266	1 090 356	85 301	167 725
饶　平	Raoping	309 750	754 399	40 473	239 225
揭　东	Jiedong	368 244	2 104 412	103 201	253 483
惠　来	Huilai	428 683	947 908	44 005	233 808
广　西	**Guangxi**	**748 788**	**751 232**	**126 307**	**422 578**
合　浦	Hepu	641 470	497 220	47 153	278 188
东　兴	Dongxing	107 318	254 012	79 154	144 390
海　南	**Hainan**	**5 078 193**	**3 396 378**	**1 074 891**	**3 265 674**
琼　海	Qionghai	604 754	249 575	153 406	321 762
儋　州	Danzhou	899 806	256 525	93 070	446 742
文　昌	Wenchang	622 657	381 629	104 467	353 169
万　宁	Wanning	408 925	361 523	91 577	289 490
东　方	Dongfang	304 312	584 452	83 115	315 797
澄　迈	Chengmai	554 732	750 819	157 343	366 628
临　高	Lingao	739 145	90 344	41 268	231 741
昌　江	Changjiang	194 398	466 325	86 275	240 160
乐　东	Ledong	446 721	78 728	58 688	332 670
陵　水	Lingshui	302 743	176 458	205 682	367 515

注：本表各省数据为合计数。

Note: The data for the provinces are the totals.

10-7　沿海地区财政收支
Financial Revenue and Expenditure by Coastal Regions

单位：亿元　　　　(100 million yuan)

地　区 Region	公共财政预算收入 Public Budgetary Revenue	公共财政预算支出 Public Budgetary Revenue
合　计 Total	**37 752.87**	**51 602.20**
天　津 Tianjin	2 079.07	2 549.21
河　北 Hebei	2 295.62	4 409.58
辽　宁 Liaoning	3 343.81	5 197.42
上　海 Shanghai	4 109.51	4 528.61
江　苏 Jiangsu	6 568.46	7 798.47
浙　江 Zhejiang	3 796.92	4 730.47
福　建 Fujian	2 119.45	3 068.80
山　东 Shandong	4 559.95	6 688.80
广　东 Guangdong	7 081.47	8 411.00
广　西 Guangxi	1 317.60	3 208.67
海　南 Hainan	481.01	1 011.17

10-8 沿海地区教育基本情况
Basic Conditions of Education by Coastal Regions

地 区 Region	高等学校数 （所） Institutions of Higher Education (unit)	本、专科在校学生数 （人） Number of Students at School in Undergraduate or Specialized Courses (person)	本、专科毕（结）业生数 （人） Number of Students Graduated in Undergraduate or Specialized Courses (person)
全国总计 National Total	**2 491**	**24 680 726**	**6 387 210**
天 津 Tianjin	55	489 919	120 996
河 北 Hebei	118	1 174 374	334 278
辽 宁 Liaoning	115	968 034	241 049
上 海 Shanghai	68	504 771	133 794
江 苏 Jiangsu	156	1 684 455	473 843
浙 江 Zhejiang	102	959 629	244 860
福 建 Fujian	87	730 510	187 230
山 东 Shandong	139	1 698 545	475 858
广 东 Guangdong	138	1 709 881	412 315
广 西 Guangxi	70	656 127	169 543
海 南 Hainan	17	172 143	43 804

10-9 沿海地区卫生基本情况
Basic Conditions of Public Health by Coastal Regions

地 区 Region	卫生机构数 （个） Health Institutions (unit)	医疗机构床位数 （万张） Total Beds (10 000 beds)	卫生机构人员 （人） Number of Employed Personnel in Health Institutions (person)
全国总计 National Total	**974 398**	**618.19**	**9 790 483**
天 津 Tianjin	4 689	5.77	106 527
河 北 Hebei	78 485	30.35	492 012
辽 宁 Liaoning	35 612	24.19	338 443
上 海 Shanghai	4 929	11.43	192 333
江 苏 Jiangsu	30 998	36.83	551 113
浙 江 Zhejiang	30 063	23.01	427 072
福 建 Fujian	28 175	15.61	261 784
山 东 Shandong	75 426	48.97	819 348
广 东 Guangdong	47 835	37.84	708 036
广 西 Guangxi	33 943	18.72	334 849
海 南 Hainan	5 011	3.21	63 468

10-10　分地区电力消费量
Electricity Consumption by Region

单位：亿千瓦·小时　(100 million kW·h)

地　区 Region	2012	2013
天　津 Tianjin	722.48	774.49
河　北 Hebei	3 077.99	3 251.19
辽　宁 Liaoning	1 899.88	2 008.46
上　海 Shanghai	1 353.45	1 410.60
江　苏 Jiangsu	4 580.90	4 956.62
浙　江 Zhejiang	3 210.55	3 453.05
福　建 Fujian	1 579.50	1 700.73
山　东 Shandong	3 794.55	4 083.12
广　东 Guangdong	4 619.41	4 830.13
广　西 Guangxi	1 153.85	1 237.74
海　南 Hainan	210.85	232.02

10-11 沿海地区用水情况
Water Use of Coastal Regions

地　区 Region	全年供水总量 （万立方米） Total Annual Volume of Water Supply ($10\ 000\ m^3$)	人均用水量 （立方米/人） Per Capita Water Use (m^3/person)
全国总计 National Total	**6 183.4**	**455.5**
天　津 Tianjin	23.8	164.7
河　北 Hebei	191.3	261.7
辽　宁 Liaoning	142.1	323.8
上　海 Shanghai	123.2	513.9
江　苏 Jiangsu	576.7	727.3
浙　江 Zhejiang	198.3	361.4
福　建 Fujian	204.8	544.6
山　东 Shandong	217.9	224.5
广　东 Guangdong	443.2	417.3
广　西 Guangxi	308.2	655.6
海　南 Hainan	43.2	484.5

10-12 沿海地区全社会固定资产投资
Total Investment in Fixed Assets by the Whole Society of Coastal Regions

单位：亿元 (100 million yuan)

地　区 Region	全社会固定资产投资 Total Investment in Fixed Assets in the Whole Country	#农、林、牧、渔业 Agriculture, Forestry, Animal Husbandry and Fishery	#交通运输、仓储和邮政业 Transport, Storage and Post
全国总计 National Total	**446 294.1**	**13 478.8**	**36 790.1**
天　津 Tianjin	9 130.2	226.1	603.1
河　北 Hebei	23 194.2	901.3	2 123.6
辽　宁 Liaoning	25 107.7	575.5	1 582.4
上　海 Shanghai	5 647.8	18.4	499.0
江　苏 Jiangsu	36 373.3	253.3	1 685.9
浙　江 Zhejiang	20 782.1	269.4	1 454.7
福　建 Fujian	15 327.4	323.4	1 572.6
山　东 Shandong	36 789.1	1 064.4	2 055.8
广　东 Guangdong	22 308.4	398.5	2 444.4
广　西 Guangxi	11 907.7	562.8	1 121.2
海　南 Hainan	2 697.9	26.5	278.7

10-13 沿海地区按经营单位所在地分货物进出口总额 Total Value of Imports and Exports by Location of Importers/Exporters of Coastal Regions

单位：万美元 (10 000 USD)

地 区 Region	进出口 Total	出口 Exports	进口 Imports
全国总计 National Total	**415 899 347**	**220 900 400**	**194 998 947**
天 津 Tianjin	12 850 179	4 900 494	7 949 685
河 北 Hebei	5 491 157	3 096 061	2 395 096
辽 宁 Liaoning	11 447 819	6 452 201	4 995 618
上 海 Shanghai	44 126 822	20 418 003	23 708 819
江 苏 Jiangsu	55 080 227	32 880 175	22 200 052
浙 江 Zhejiang	33 578 871	24 874 624	8 704 246
福 建 Fujian	16 932 090	10 647 442	6 284 648
山 东 Shandong	26 653 153	13 419 013	13 234 141
广 东 Guangdong	109 158 144	63 636 385	45 521 759
广 西 Guangxi	3 282 750	1 869 326	1 413 424
海 南 Hainan	1 498 543	370 649	1 127 895

10-14 沿海地区城镇居民平均每人全年家庭收入和消费性支出 Per Capita Annual Income and Consumption Expenditure of Urban Households by Coastal Regions

单位：元 (yuan)

地 区 Region	可支配收入 Disposable Income	现金可支配收入 Cash Disposable Income	消费支出 Expenses on Consumption	现金消费支出 Cash Consumption Expenditure
全国平均水平 National Average	**18 310.8**	**17 114.6**	**13 220.4**	**10 917.4**
天 津 Tianjin	26 359.2	24 700.4	20 418.7	17 064.1
河 北 Hebei	15 189.6	14 325.9	10 872.2	9 177.1
辽 宁 Liaoning	20 817.8	19 862.4	14 950.2	12 689.2
上 海 Shanghai	42 173.6	35 418.7	30 399.9	21 417.4
江 苏 Jiangsu	24 775.5	23 188.2	17 925.8	14 634.1
浙 江 Zhejiang	29 775.0	28 081.3	20 610.1	16 543.2
福 建 Fujian	21 217.9	20 097.3	16 176.6	13 158.6
山 东 Shandong	19 008.3	18 239.2	11 896.8	10 078.2
广 东 Guangdong	23 420.7	22 129.8	17 421.0	14 811.3
广 西 Guangxi	14 082.3	13 157.7	9 596.5	7 753.1
海 南 Hainan	15 733.3	14 953.3	11 192.9	9 330.7

10-15 沿海地区农村居民家庭人均纯收入和生活消费支出
Per Capita Annual Net Income and Consumption Expenditure of Rural Households by Coastal Regions

单位：元 (yuan)

地 区 Region	纯收入 Net Income	生活消费支出 Consumption Expenditure
全国平均水平 National Average	**8 895.9**	**6 625.5**
天 津 Tianjin	15 841.0	10 155.0
河 北 Hebei	9 101.9	6 134.1
辽 宁 Liaoning	10 522.7	7 159.0
上 海 Shanghai	19 595.0	14 234.7
江 苏 Jiangsu	13 597.8	9 909.8
浙 江 Zhejiang	16 106.0	11 760.2
福 建 Fujian	11 184.2	8 151.2
山 东 Shandong	10 619.9	7 392.7
广 东 Guangdong	11 669.3	8 343.5
广 西 Guangxi	6 790.9	5 205.6
海 南 Hainan	8 342.6	5 465.6

10-16 沿海地区年末人口数
Year-end Population by Coastal Regions

单位：万人 (10 000 persons)

地 区 Region	2006	2007	2008	2009	2010	2011	2012	2013
全国总计 National Total	**131 448**	**132 129**	**132 802**	**133 450**	**134 091**	**134 735**	**135 404**	**136 072**
天 津 Tianjin	1 705	1 115	1 176	1 228	1 299	1 355	1 413	1 472
河 北 Hebei	6 898	6 943	6 989	7 034	7 194	7 241	7 288	7 333
辽 宁 Liaoning	4 271	4 298	4 315	4 341	4 375	4 383	4 389	4 390
上 海 Shanghai	1 964	2 064	2 141	2 210	2 303	2 347	2 380	2 415
江 苏 Jiangsu	7 656	7 723	7 762	7 810	7 869	7 899	7 920	7 939
浙 江 Zhejiang	5 072	5 155	5 212	5 276	5 447	5 463	5 477	5 498
福 建 Fujian	3 585	3 612	3 639	3 666	3 693	3 720	3 748	3 774
山 东 Shandong	9 309	9 367	9 417	9 470	9 588	9 637	9 685	9 733
广 东 Guangdong	9 442	9 660	9 893	10 130	10 441	10 505	10 594	10 644
广 西 Guangxi	4 719	4 768	4 816	4 856	4 610	4 645	4 682	4 719
海 南 Hainan	836	845	854	864	869	877	887	895

注:2010年数据为当年人口普查数据推算数；其余年份数据为年度人口抽样调查推算数据。各地区数据为常住人口口径。

Note: Data of 2010 are the census year estimates; the rest are the estimates from the annual national sample survey on population changes. Data by region are usual residents.

10-17 沿海城市人口情况（2012年）
Population of Coastal Cities, 2012

单位：万人 (10 000 persons)

沿海城市 Coastal City		常住人口 Permanent Population	年底总人口 Total Population by the End of the Year
合　计	**Total**	**28 889.8**	**24 522.1**
天　津	**Tianjin**	**1 413.0**	**993.3**
河　北	**Hebei**	**1 793.5**	**1 777.4**
唐　山	Tangshan	766.9	741.8
秦皇岛	Qinhuangdao	302.2	291.2
沧　州	Cangzhou	724.4	744.4
辽　宁	**Liaoning**	**1 889.3**	**1 782.5**
大　连	Dalian	689.2	590.3
丹　东	Dandong	243.1	240.5
锦　州	Jinzhou	309.7	307.8
营　口	Yingkou	244.2	235.1
盘　锦	Panjin	143.5	128.8
葫芦岛	Huludao	259.6	280.0
上　海	**Shanghai**	**2 380.6**	**1 427.0**
江　苏	**Jiangsu**	**1 892.0**	**2 098.6**
南　通	Nantong	729.7	765.2
连云港	Lianyungang	440.7	511.0
盐　城	Yancheng	721.6	822.4

10-17 续表1 continued

沿海城市	Coastal City	常住人口 Permanent Population	年底总人口 Total Population by the End of the Year
浙　江	**Zhejiang**	**4 222.9**	**3 551.8**
杭　州	Hangzhou	880.2	700.5
宁　波	Ningbo	763.9	577.7
温　州	Wenzhou	915.6	800.2
嘉　兴	Jiaxing	454.4	344.5
绍　兴	Shaoxing	494.3	440.8
舟　山	Zhoushan	114.0	97.2
台　州	Taizhou	600.5	590.9
福　建	**Fujian**	**2 978.0**	**2 693.5**
福　州	Fuzhou	727.0	655.3
厦　门	Xiamen	367.0	190.9
莆　田	Putian	281.0	329.3
泉　州	Quanzhou	829.0	693.2
漳　州	Zhangzhou	490.0	482.5
宁　德	Ningde	284.0	342.3
山　东	**Shandong**	**3 656.2**	**3 406.9**
青　岛	Qingdao	886.9	769.6
东　营	Dongying	207.3	185.5
烟　台	Yantai	698.3	650.3
潍　坊	Weifang	921.6	878.9
威　海	Weihai	279.8	253.6
日　照	Rizhao	283.4	288.1
滨　州	Binzhou	378.9	380.9

10-17 续表2 continued

沿海城市	Coastal City	常住人口 Permanent Population	年底总人口 Total Population by the End of the Year
广　东	**Guangdong**	**7 819.3**	**5 935.3**
广　州	Guangzhou	1 283.9	822.3
深　圳	Shenzhen	1 054.7	299.2
珠　海	Zhuhai	158.3	106.6
汕　头	Shantou	544.8	532.9
江　门	Jiangmen	448.3	391.8
湛　江	Zhanjiang	710.9	785.2
茂　名	Maoming	596.8	748.9
惠　州	Huizhou	467.4	341.9
汕　尾	Shanwei	296.9	347.2
阳　江	Yangjiang	247.0	282.5
东　莞	Dongguan	829.2	186.1
中　山	Zhongshan	315.5	152.0
潮　州	Chaozhou	270.0	264.8
揭　阳	Jieyang	595.6	673.9
广　西	**Guangxi**	**558.7**	**636.9**
北　海	Beihai	156.7	164.4
防城港	Fangchenggang	88.7	87.3
钦　州	Qinzhou	313.3	385.2
海　南	**Hainan**	**286.3**	**218.9**
海　口	Haikou	214.1	161.6
三　亚	Sanya	72.2	57.3

注：本表各省数据为合计数。

Note: The data for the provinces are the totals.

10-18 沿海县人口情况（2012年）
Population of Coastal Counties, 2012

单位：万人 (10 000 persons)

沿海县	Coastal County	年末总人口 Total Population by the End of the Year	乡村人口 Rural Population
合 计	**Total**	**8 579.0**	**7 061.2**
河 北	**Hebei**	**282.4**	**240.7**
滦 南	Luannan	57.1	52.9
乐 亭	Leting	49.3	43.1
唐 海	Tanghai		
昌 黎	Changli	56.2	46.2
抚 宁	Funing	49.6	41.4
黄 骅	Huanghua	46.5	38.4
海 兴	Haixing	23.7	18.7
辽 宁	**Liaoning**	**665.5**	**518.6**
长 海	Changhai	7.3	4.2
瓦房店	Wafangdian	100.2	72.9
普兰店	Pulandian	93.1	57.9
庄 河	Zhuanghe	90.5	70.9
东 港	Donggang	60.7	49.2
凌 海	Linghai	52.7	44.7
盖 州	Gaizhou	71.9	61.9
大 洼	Dawa	40.2	32.2
盘 山	Panshan	29.9	28.1
绥 中	Suizhong	64.2	55.2
兴 城	Xingcheng	54.8	41.4

10-18 续表1 continued

沿海县 Coastal County		年末总人口 Total Population by the End of the Year	乡村人口 Rural Population
江　苏	**Jiangsu**	**1 289.6**	**1 009.6**
海　安	Hai'an	93.9	71.1
如　东	Rudong	104.6	86.3
启　东	Qidong	112.4	90.8
海　门	Haimen	100.0	79.3
赣　榆	Ganyu	115.6	88.7
东　海	Donghai	118.0	93.2
灌　云	Guanyun	102.0	81.0
灌　南	Guannan	78.7	61.0
响　水	Xiangshui	61.5	46.1
滨　海	Binhai	120.1	93.5
射　阳	Sheyang	96.7	74.3
东　台	Dongtai	113.6	90.7
大　丰	Dafeng	72.5	53.6
浙　江	**Zhejiang**	**1 486.0**	**1 325.1**
象　山	Xiangshan	54.0	39.9
宁　海	Ninghai	61.6	50.9
余　姚	Yuyao	83.5	70.8
慈　溪	Cixi	104.2	119.8
奉　化	Fenghua	48.4	38.9
洞　头	Dongtou	13.1	8.4
平　阳	Pingyang	87.3	68.1
苍　南	Cangnan	130.0	100.8
瑞　安	Rui'an	121.6	115.2
乐　清	Yueqing	127.2	124.2
海　盐	Haiyan	37.6	35.2
海　宁	Haining	66.6	52.9
平　湖	Pinghu	48.9	34.5

10-18 续表2 continued

沿海县 Coastal County		年末总人口 Total Population by the End of the Year	乡村人口 Rural Population
绍　兴	Shaoxing	72.7	84.8
上　虞	Shangyu	77.8	64.6
岱　山	Daishan	19.0	14.3
嵊　泗	Shengsi	7.9	4.5
玉　环	Yuhuan	42.6	61.2
三　门	Sanmen	43.5	34.5
温　岭	Wenling	120.6	112.1
临　海	Linhai	117.9	89.5
福　建	**Fujian**	**1 300.9**	**1 128.3**
连　江	Lianjiang	64.1	56.9
罗　源	Luoyuan	25.8	23.0
平　潭	Pingtan	41.3	36.6
福　清	Fuqing	129.4	114.9
长　乐	Changle	69.4	62.7
仙　游	Xianyou	109.9	98.1
惠　安	Hui'an	97.3	89.3
石　狮	Shishi	31.8	24.7
晋　江	Jinjiang	107.4	115.0
南　安	Nan'an	151.7	113.7
云　霄	Yunxiao	44.1	35.1
漳　浦	Zhangpu	86.6	79.6
诏　安	Zhao'an	61.1	56.6
东　山	Dongshan	20.9	14.0
龙　海	Longhai	83.1	71.1
霞　浦	Xiapu	53.5	43.7
福　安	Fu'an	65.1	45.3
福　鼎	Fuding	58.4	48.0

10-18 续表3 continued

沿海县 Coastal County		年末总人口 Total Population by the End of the Year	乡村人口 Rural Population
山　东	**Shandong**	**1 140.2**	**934.1**
胶　州	Jiaozhou	80.9	63.0
即　墨	Jimo	113.4	97.6
胶　南	Jiaonan		
垦　利	Kenli	22.0	18.0
利　津	Lijin	30.0	26.0
广　饶	Guangrao	50.1	43.5
长　岛	Changdao	4.3	2.7
龙　口	Longkou	63.5	49.1
莱　阳	Laiyang	86.8	75.1
莱　州	Laizhou	85.6	68.2
蓬　莱	Penglai	45.0	38.0
招　远	Zhaoyuan	56.8	44.9
海　阳	Haiyang	66.0	58.6
寿　光	Shouguang	105.1	88.9
昌　邑	Changyi	58.3	49.9
文　登	Wendeng	63.6	47.1
荣　成	Rongcheng	67.0	44.1
乳　山	Rushan	56.7	47.2
无　棣	Wudi	46.0	37.3
沾　化	Zhanhua	39.1	34.9
广　东	**Guangdong**	**1 752.3**	**1 431.8**
南　澳	Nan'ao	7.4	5.7
台　山	Taishan	98.3	84.7
恩　平	Enping	50.0	33.0
遂　溪	Suixi	105.0	92.6
徐　闻	Xuwen	73.1	61.0

10-18 续表4 continued

沿海县 Coastal County		年末总人口 Total Population by the End of the Year	乡村人口 Rural Population
廉　江	Lianjiang	171.3	134.8
雷　州	Leizhou	168.7	156.9
吴　川	Wuchuan	111.1	86.0
电　白	Dianbai	147.1	119.1
惠　东	Huidong	88.2	54.5
海　丰	Haifeng	81.5	68.3
陆　丰	Lufeng	180.9	113.0
阳　西	Yangxi	51.4	41.7
阳　东	Yangdong	48.5	46.3
饶　平	Raoping	103.4	89.8
揭　东	Jiedong	131.4	121.7
惠　来	Huilai	135.0	122.7
广　西	**Guangxi**	**119.1**	**73.0**
东　兴	Dongxing	13.5	9.3
合　浦	Hepu	105.6	63.7
海　南	**Hainan**	**543.0**	**400.0**
琼　海	Qionghai	50.0	37.0
儋　州	Danzhou	98.0	64.0
文　昌	Wenchang	59.0	47.0
万　宁	Wanning	62.0	44.0
东　方	Dongfang	48.0	36.0
澄　迈	Chengmai	57.0	43.0
临　高	Lingao	51.0	38.0
昌　江	Changjiang	26.0	17.0
乐　东	Ledong	54.0	45.0
陵　水	Lingshui	38.0	29.0

注：本表各省数据为合计数。

Note: The data for the provinces are the totals.

10-19 沿海地区城镇单位就业人员情况
Employed Persons in Urban Units by Coastal Regions

单位：万人 (10 000 persons)

地　区 Region	2012	2013
全国总计 National Total	**15 236.4**	**18 108.4**
天　津 Tianjin	289.1	302.4
河　北 Hebei	619.9	653.4
辽　宁 Liaoning	598.7	689.1
上　海 Shanghai	555.7	618.8
江　苏 Jiangsu	830.9	1 503.3
浙　江 Zhejiang	1 070.1	1 071.6
福　建 Fujian	637.9	644.0
山　东 Shandong	1 110.2	1 290.6
广　东 Guangdong	1 304.0	1 967.0
广　西 Guangxi	358.0	403.0
海　南 Hainan	90.1	98.8

10-20 沿海城市就业人员情况（2012年）
Number of Employed Persons by Coastal Cities, 2012

单位：万人 (10 000 persons)

沿海城市 Coastal City		就业人员 Number of Employed Persons	城镇就业人员 Urban Employed Persons
合 计	**Total**	**15 354.30**	**3 694.40**
天 津	**Tianjin**		
河 北	**Hebei**	**1 033.30**	**181.50**
唐 山	Tangshan	449.00	95.60
秦皇岛	Qinhuangdao	167.00	33.50
沧 州	Cangzhou	417.30	52.40
辽 宁	**Liaoning**	**1 124.40**	**262.90**
大 连	Dalian	441.90	111.50
丹 东	Dandong	124.90	26.90
锦 州	Jinzhou	164.80	29.20
营 口	Yingkou	155.00	28.00
盘 锦	Panjin	100.90	46.90
葫芦岛	Huludao	136.90	20.40
上 海	**Shanghai**		
江 苏	**Jiangsu**	**1 165.80**	**157.80**
南 通	Nantong	468.90	68.50
连云港	Lianyungang	249.20	35.50
盐 城	Yancheng	447.70	53.80
浙 江	**Zhejiang**	**2 857.10**	**892.50**
杭 州	Hangzhou	644.40	281.90
宁 波	Ningbo	501.60	174.70
温 州	Wenzhou	577.90	111.50
嘉 兴	Jiaxing	327.10	79.20
绍 兴	Shaoxing	343.90	130.70
舟 山	Zhoushan	72.90	17.70
台 州	Taizhou	389.30	96.80
福 建	**Fujian**	**1 998.00**	**537.10**
福 州	Fuzhou	451.90	143.80
厦 门	Xiamen	276.40	118.00
莆 田	Putian	203.60	36.90

10-20 续表 continued

沿海城市 Coastal City		就业人员 Number of Employed Persons	城镇就业人员 Urban Employed Persons
泉　州	Quanzhou	580.90	163.60
漳　州	Zhangzhou	294.80	48.60
宁　德	Ningde	190.40	26.20
山　东	**Shandong**	**2 250.40**	**486.10**
青　岛	Qingdao	548.50	129.00
东　营	Dongying	128.80	47.10
烟　台	Yantai	436.10	104.70
潍　坊	Weifang	519.70	83.70
威　海	Weihai	174.80	56.80
日　照	Rizhao	188.40	21.80
滨　州	Binzhou	254.10	43.00
广　东	**Guangdong**	**4 493.10**	**1 080.50**
广　州	Guangzhou	751.30	326.80
深　圳	Shenzhen	771.20	280.00
珠　海	Zhuhai	104.90	65.40
汕　头	Shantou	239.00	53.00
江　门	Jiangmen	248.30	53.30
湛　江	Zhanjiang	331.60	44.40
茂　名	Maoming	278.30	38.30
惠　州	Huizhou	270.00	89.00
汕　尾	Shanwei	119.40	17.50
阳　江	Yangjiang	132.00	18.40
东　莞	Dongguan	631.40	25.30
中　山	Zhongshan	208.80	33.00
潮　州	Chaozhou	135.00	13.10
揭　阳	Jieyang	271.90	23.00
广　西	**Guangxi**	**160.10**	**41.30**
北　海	Beihai	84.10	14.30
防城港	Fangchenggang	59.40	10.40
钦　州	Qinzhou	16.60	16.60
海　南	**Hainan**	**272.10**	**54.70**
海　口	Haikou	136.60	45.00
三　亚	Sanya	135.50	9.70

注：本表各省数据为合计数。

Note: The data for the provinces are the totals.

10-21 沿海县就业人员情况（2012年）
Number of Employed Persons of Coastal Counties, 2012

单位：人 (person)

沿海县 Coastal County		年末单位从业人员数 Employed Persons by the End of the year	乡村从业人员数 Rural Employees
合　计	**Total**	**8 889 038**	**38 764 288**
河　北	**Hebei**	**165 101**	**1 326 931**
丰　南	Fengnan		
滦　南	Luannan	31 338	287 795
乐　亭	Leting	26 809	262 405
唐　海	Tanghai		
昌　黎	Changli	22 562	268 011
抚　宁	Funing	37 987	222 734
黄　骅	Huanghua	33 272	182 863
海　兴	Haixing	13 133	103 123
辽　宁	**Liaoning**	**568 894**	**2 723 344**
长　海	Changhai	8 319	22 277
瓦房店	Wafangdian	61 248	366 554
普兰店	Pulandian	62 056	281 220
庄　河	Zhuanghe	45 320	370 836
东　港	Donggang	44 122	265 819
凌　海	Linghai	31 526	244 357
盖　州	Gaizhou	22 011	333 262
大　洼	Dawa	179 853	187 869
盘　山	Panshan	63 830	162 473
绥　中	Suizhong	22 763	285 278
兴　城	Xingcheng	27 846	203 399

10-21 续表1 continued

沿海县 Coastal County		年末单位从业人员数 Employed Persons by the End of the year	乡村从业人员数 Rural Employees
江 苏	**Jiangsu**	**702 113**	**5 257 500**
海 安	Hai'an	76 021	381 600
如 东	Rudong	69 855	474 800
启 东	Qidong	72 541	518 700
海 门	Haimen	68 821	496 700
赣 榆	Ganyu	40 194	425 900
东 海	Donghai	39 801	451 600
灌 云	Guanyun	33 224	375 200
灌 南	Guannan	35 709	320 500
响 水	Xiangshui	33 378	212 300
滨 海	Binhai	39 591	449 600
射 阳	Sheyang	57 486	349 100
东 台	Dongtai	70 007	488 100
大 丰	Dafeng	65 485	313 400
浙 江	**Zhejiang**	**2 577 078**	**8 184 700**
象 山	Xiangshan	313 822	252 400
宁 海	Ninghai	70 739	323 600
余 姚	Yuyao	152 279	433 200
慈 溪	Cixi	159 237	778 800
奉 化	Fenghua	58 439	268 000
洞 头	Dongtou	7 846	54 300
平 阳	Pingyang	79 865	418 200
苍 南	Cangnan	96 147	622 400
瑞 安	Rui'an	106 500	661 500
乐 清	Yueqing	218 267	711 800
海 盐	Haiyan	63 918	219 200
海 宁	Haining	125 373	315 800
平 湖	Pinghu	130 828	217 500

10-21 续表2 continued

沿海县 Coastal County		年末单位从业人员数 Employed Persons by the End of the year	乡村从业人员数 Rural Employees
绍　兴	Shaoxing	329 598	523 800
上　虞	Shangyu	215 846	378 500
岱　山	Daishan	21 500	85 000
嵊　泗	Shengsi	11 500	27 100
玉　环	Yuhuan	110 263	431 900
三　门	Sanmen	35 888	219 900
温　岭	Wenling	109 165	662 200
临　海	Linhai	160 058	579 600
福　建	**Fujian**	**1 866 219**	**6 070 296**
连　江	Lianjiang	45 954	295 708
罗　源	Luoyuan	18 625	103 288
平　潭	Pingtan	22 082	186 624
福　清	Fuqing	209 797	545 762
长　乐	Changle	89 034	282 633
仙　游	Xianyou	51 262	500 200
惠　安	Hui'an	303 800	480 953
石　狮	Shishi	148 713	134 908
晋　江	Jinjiang	569 446	696 108
南　安	Nan'an	114 141	760 591
云　霄	Yunxiao	28 414	171 588
漳　浦	Zhangpu	59 217	451 759
诏　安	Zhao'an	37 721	328 853
东　山	Dongshan	19 168	80 568
龙　海	Longhai	76 744	397 400
霞　浦	Xiapu	19 039	220 018
福　安	Fu'an	28 928	170 706
福　鼎	Fuding	24 134	262 629

10-21 续表3 continued

沿海县 Coastal County		年末单位从业人员数 Employed Persons by the End of the year	乡村从业人员数 Rural Employees
山　东	**Shandong**	**1 477 325**	**5 204 026**
胶　州	Jiaozhou	178 930	361 050
即　墨	Jimo	170 298	558 564
胶　南	Jiaonan		
垦　利	Kenli	37 361	90 147
利　津	Lijin	31 545	147 493
广　饶	Guangrao	72 158	274 618
长　岛	Changdao	5 324	13 227
龙　口	Longkou	98 734	270 957
莱　阳	Laiyang	84 641	421 876
莱　州	Laizhou	78 129	378 437
蓬　莱	Penglai	90 213	200 370
招　远	Zhaoyuan	86 051	215 296
海　阳	Haiyang	38 847	356 797
寿　光	Shouguang	97 326	422 831
昌　邑	Changyi	39 516	267 882
文　登	Wendeng	104 218	274 326
荣　成	Rongcheng	149 389	230 086
乳　山	Rushan	56 633	266 201
无　棣	Wudi	34 065	245 182
沾　化	Zhanhua	23 947	208 686
广　东	**Guangdong**	**986 283**	**7 422 887**
南　澳	Nan'ao	5 120	24 260
台　山	Taishan	56 258	513 390
恩　平	Enping	29 533	187 701
遂　溪	Suixi	40 532	458 851
徐　闻	Xuwen	39 906	320 548

10-21 续表4 continued

沿海县 Coastal County		年末单位从业人员数 Employed Persons by the End of the year	乡村从业人员数 Rural Employees
廉　江	Lianjiang	61 795	704 920
雷　州	Leizhou	58 111	718 808
吴　川	Wuchuan	50 679	545 004
电　白	Dianbai	72 773	557 751
惠　东	Huidong	66 082	378 090
海　丰	Haifeng	41 119	402 398
陆　丰	Lufeng	51 661	643 184
阳　西	Yangxi	264 268	200 854
阳　东	Yangdong	30 043	262 203
饶　平	Raoping	30 393	435 177
揭　东	Jiedong	49 532	606 094
惠　来	Huilai	38 478	463 654
广　西	**Guangxi**	**56 669**	**402 400**
东　兴	Dongxing	10 293	53 500
合　浦	Hepu	46 376	348 900
海　南	**Hainan**	**489 356**	**2 172 204**
琼　海	Qionghai	52 628	193 807
儋　州	Danzhou	40 688	323 529
文　昌	Wenchang	27 302	279 451
万　宁	Wanning	90 114	214 265
东　方	Dongfang	24 802	183 749
澄　迈	Chengmai	120 363	249 144
临　高	Lingao	16 401	224 930
昌　江	Changjiang	20 243	89 438
乐　东	Ledong	80 428	247 858
陵　水	Lingshui	16 387	166 033

注：本表各省数据为合计数，缺少福建金门县数据。

Note: The data for the provinces are the totals, and lacking the data for Jinmen of Fujian Province.

主要统计指标解释

1. 国内(或地区)生产总值 指一个国家（或地区）所有常驻单位在一定时期内生产活动的最终成果。国内生产总值有三种表现形态，即价值形态、收入形态和产品形态。从价值形态看，它是所有常驻单位在一定时期内生产的全部货物和服务价值超过同期中间投入的全部非固定资产货物和服务价值的差额，即所有常驻单位的增加值之和；从收入形态看，它是所有常驻单位在一定时期内创造并分配给常驻单位和非常驻单位的初次收入分配之和；从产品形态看，它是所有常驻单位在一定时期内最终使用的货物和服务价值与货物和服务净出口价值之和。在实际核算中，国内生产总值有三种计算方法，即生产法、收入法和支出法。三种方法分别从不同的方面反映国内生产总值及其构成。

2. 三次产业 是根据社会生产活动历史发展的顺序对产业结构的划分，产品直接取自自然界的部门称为第一产业，对初级产品进行再加工的部门称为第二产业，为生产和消费提供各种服务的部门称为第三产业。它是世界上较为通用的产业结构分类，但各国的划分不尽一致。我国的三次产业划分是：

第一产业：是指农、林、牧、渔业。

第二产业：是指采矿业，制造业，电力、燃气及水的生产和供应业，建筑业。

第三产业：是指除第一、第二产业以外的其他行业。第三产业包括：交通运输、仓储和邮政业，信息传输、计算机服务和软件业，批发和零售业，住宿和餐饮业，金融业，房地产业，租赁和商务服务业，科学研究、技术服务和地质勘查业，水利、环境和公共设施管理业，居民服务和其他服务业，教育，卫生、社会保障和社会福利业，文化、体育和娱乐业，公共管理和社会组织，国际组织。

3. 增加值 是指各行各业生产经营和劳务活动的最终成果，采用生产法和收入法两种方法计算。

生产法 是从货物和服务活动在生产过程中形成的总产品入手，剔除生产过程中投入的中间产品价值，得到新增价值的方法。

收入法 又称分配法。按收入法计算国内生产总值是从生产过程创造的收入的角度对常驻单位的生产活动成果进行核算；按照此法计算，增加值由劳动者报酬、固定资产折旧、生产税净额和营业盈余四个部分组成。

4. 年末总人口 是指每年 12 月 31 日 24 时一定地区范围内的有生命的个人的人口总和。

5. 财政收入 指国家财政参与社会产品分配所取得的收入，是实现国家职能的财力保证。财政收入所包括的内容几经变化，目前主要包括：

（1）各项税收 包括增值税、营业税、消费税、土地增值税、城市维护建设税、资源税、城市土地使用税、企业所得税、个人所得税、关税、证券交易印花税、车辆购置税、农牧业税和耕地占用税等。

（2）专项收入 包括排污费收入、城市水资源费收入、矿产资源补偿费收入、教育费附加收入等。

（3）其他收入 包括利息收入、基本建设贷款归还收入、基本建设收入、捐赠收入等。

（4）国有企业亏损补贴　此项为负收入，冲减财政收入。主要包括对工业企业、商业企业、粮食企业的补贴。

6. 财政支出　国家财政将筹集起来的资金进行分配使用，以满足经济建设和各项事业的需要。

7. 基本建设支出　指按国家有关规定，属于基本建设范围内的基本建设有偿使用、拨款、资本金支出以及经国家批准对专项和政策性基建投资贷款，在部门的基建投资额中统筹支付的贴息支出。

8. 普通高等学校　指按照国家规定的设置标准和审批程序批准举办的，通过全国普通高等学校统一招生考试，招收高中毕业生为主要培养对象，实施高等教育的全日制大学、独立设置的学院和高等专科学校、高等职业学校和其他机构。

大学、独立设置的学院主要实施本科层次以上教育，高等专科学校、高等职业学校实施专科层次教育，其他机构是承担国家普通招生计划任务不计校数的机构。包括普通高等学校分校和批准筹建的普通高等学校等。

9. 卫生机构　包括医疗机构、疾病预防控制中心(防疫站)、采供血机构、卫生监督及监测(检验)机构、医学科研和在职培训机构、健康教育所等。

10. 医疗机构　包括医院、社区卫生服务中心(站)、疗养院、卫生院、门诊部、诊所(卫生所、医务室)、妇幼保健院(所、站)、专科疾病防治院(所、站)、急救中心(站)和临床检验中心。医疗机构分为非赢利性医疗机构和赢利性医疗机构。

11. 单位国内生产总值能耗　指一定时期内，一个国家或地区每生产一个单位的国内生产总值所消耗的能源。计算公式为

$$\text{单位国内生产总值能源}=\frac{\text{能源消费总量}}{\text{国内生产总值}}$$

12. 单位国内生产总值电耗　指一定时期内，一个国家或地区每生产一个单位的国内生产总值所消耗的电力。计算公式为

$$\text{单位国内生产总值电耗}=\frac{\text{全社会用电量}}{\text{国内生产总值}}$$

13. 单位工业增加值能耗　指一定时期内，一个国家或地区每生产一个单位的工业增加值所消耗的能源。计算公式为

$$\text{单位工业增加值能耗}=\frac{\text{工业能源消费总量}}{\text{工业增加值}}$$

14. 用水总量　指分配给各类用户的包括输水损失在内的毛用水量之和，不包括海水直接利用量。

15. 全社会固定资产投资　是以货币形式表现的在一定时期内全社会建造和购置固定资产的工作量以及与此有关的费用的总称。该指标是反映固定资产投资规模、结构和发展速度的综合性指标，又是观察工程进度和考核投资效果的重要依据。全社会固定资产投资按登记注册类型可分为国有、集体、个体、联营、股份制、外商、港澳台商、其他等。

16. 进出口总额　指实际进出我国国境的货物总金额。包括对外贸易实际进出口货物，来料加工装配进出口货物，国家间、联合国及国际组织无偿援助物资和赠送品，华侨、港澳台同胞和外籍华人捐赠品，租赁期满归承租人所有的租赁货物，进料加工进出口货物，边境地方贸易及边

境地区小额贸易进出口货物(边民互市贸易除外)，中外合资企业、中外合作经营企业、外商独资经营企业进出口货物和公用物品，到、离岸价格在规定限额以上的进出口货样和广告品(无商业价值、无使用价值和免费提供出口的除外)，从保税仓库提取在中国境内销售的进口货物，以及其他进出口货物。该指标可以观察一个国家在对外贸易方面的总规模。我国规定出口货物按离岸价格统计，进口货物按到岸价格统计。

17. 商品经营单位所在地进、出口额 指在所在地海关注册登记的有进出口经营权的企业实际进、出口额。

18. 城镇家庭可支配收入 指家庭成员得到可用于最终消费支出和其它非义务性支出以及储蓄的总和，即居民家庭可以用来自由支配的收入。它是家庭总收入扣除交纳的所得税、个人交纳的社会保障支出以及记账补贴后的收入。计算公式为

可支配收入=家庭总收入-交纳所得税-个人交纳的社会保障支出-记账补贴

19. 城镇家庭消费性支出 指家庭用于日常生活的支出，包括食品、衣着、家庭设备用品及服务、医疗保健、交通和通信、娱乐教育文化服务、居住、杂项商品和服务等八大类支出。

20. 总收入 指调查期内农村住户和住户成员从各种来源渠道得到的收入总和。按收入的性质划分为工资性收入、家庭经营收入、财产性收入和转移性收入。

21. 纯收入 指农村住户当年从各个来源得到的总收入相应地扣除所发生的费用后的收入总和。计算方法

纯收入=总收入-税费支出-家庭经营费用支出-生产性固定资产折旧-赠送农村亲友支出

纯收入主要用于再生产投入和当年生活消费支出，也可用于储蓄和各种非义务性支出。“农民人均纯收入”按人口平均的纯收入水平，反映的是一个地区或一个农户农村居民的平均收入水平。

22. 人口数 指一定时点、一定地区范围内有生命的个人总和。

年度统计的年末人口数指每年 12 月 31 日 24 时的人口数。年度统计的全国人口总数内未包括香港、澳门特别行政区和台湾省以及海外华侨人数。

23. 城镇人口 城镇人口是指居住在城镇范围内的全部常住人口

24. 就业人员 指在 16 周岁及以上，从事一定社会劳动并取得劳动报酬或经营收入的人员。这一指标反映了一定时期内全部劳动力资源的实际利用情况，是研究我国基本国情国力的重要指标。

Explanatory Notes on Main Statistical Indicators

1. Gross Domestic Product (GDP) refers to the final result of the primary distribution of the income created by all the resident units of a country (or a region) during a certain period of time. Gross domestic product is expressed in three different forms, i.e. value, income, and products respectively. The form of value refers to the total value of all products and services produced by all resident units during a certain period of time minus total value of intermediate input of materials and services of the nature of non-fixed assets or the summation of the value added of all resident units: the form of income includes all the income created by all resident units and distributed primarily to all resident and non-resident units; the form of products refers to the summation of the value of the products and services finally used and the net export value of products and services by all resident units during a given period of time. In the practice of national accounting, gross domestic product is calculated with three approaches, i.e. production approach, income approach, and expenditure approach, which reflect the gross domestic product and its composition from different aspects.

2. Three Industries Industrial structure is classified according to the sequence of historical development of social productive activities. Primary industry refers to the extraction of natural resources; secondary industry involves processing of primary products; and tertiary industry provides services of various kinds for production and consumption. The above classification is universal in the world although it varies to some extent from country to country. The three industries in China are divided as follows.

Primary industry: refers to farming, forestry, sideline production and fishery.

Secondary industry: refers to such industries as mining, manufacturing, production and supply of electric power, fuel gas and water, and construction.

Tertiary industry: refers to all the other industries not included in the primary or secondary industries. It includes such industries as communications and transportation, storage and postal service; information transmission, computer service and software; wholesale and retailing; accommodation and catering; financial service; real estate; charter business and commercial affairs service; scientific research, technological service and geological survey; water conservancy, environmental and other public facilities management; residents service and other service trades; education, health, social security and social welfare; culture, sports and entertainment business as well as public administration and social organizations, and international organizations.

3. Added Value refers to the final result of production operation and labor activities of all trades and professions, which is calculated by using the methods of production and income.

Production Method: refers to the method whereby to get the newly added value by proceeding from the gross product of goods and service activities occurring in the course of production and then rejecting the value of intermediate product input in the course of production.

Income Method: is also called the distribution method. The calculation of the gross domestic product (GDP) by the income method is the accounting of the result of productive activities of permanent units from the angle of the income created in the course of production. According to this method, the added value is composed of the payment for laborers, depreciation for fixed assets, net tax on production and business surplus.

4. Total Population by the End of the Year refers to the sum of living individuals within a particular range of area at 24:00 on December 31 of each year.

5. Government Revenue refers to the income obtained by the government finance through participating in the distribution of social products. It is the financial guarantee to ensure government functioning. The contents of government revenue have changed several times. Now it includes the following main items:

(1) Various tax revenues, including value added tax, business tax, consumption tax, land value-added tax, tax on city maintenance and construction, resources tax, tax on use of urban land,

enterprise income tax, personal income tax, tariff, stamp tax on security transactions, tax on purchase of motor vehicles, tax on agriculture and animal husbandry and tax on occupancy of cultivated land, etc.

(2) Special revenues, including revenues from the fee on sewage treatment, fee on urban water resources, fee for the compensation of mineral resources and extra-charges for education, etc.

(3) Other revenues, including revenues from interest, repayment of capital construction loan, capital construction projects, and donations and grants.

(4) Subsidies for the losses of State-owned enterprises. This is an item of negative revenue, counteracting revenues and consisting of subsidies to industrial, commercial and grain purchasing and supply enterprises.

6. Government Expenditure refers to the distribution and use of the funds which the government finance has raised, so as to meet the needs of economic construction and various causes.

7. Expenditure for capital construction It refers to the non-gratuitous use of, appropriation of funds for and capital outlay on capital construction in the area of capital construction. It also covers the loans on capital construction approved by the government for special purposes or policy purposes and the expenditure with discount paid in an overall way within the amount of the funds appropriated to the departments for capital construction.

8. Regular Institutions of Higher Learning refer to educational establishments set up according to the government evaluation and approval procedures, enrolling graduates from senior secondary schools and providing higher education courses and training for senior professionals. They include full-time universities, colleges, institutions of higher professional education, institutions of higher vocational education and others.

Universities and colleges primarily provide undergraduate courses; institutions of higher professional education and institutions of higher vocational education primarily provide professional trainings; and others refer to educational establishments, which are responsible for enrolling higher education students under the State Plan but not enumerated in the total number of schools, including: branch schools of universities and colleges, and universities and colleges that have been approved and under plan for construction.

9. Health Care Institutions include: medical institutions, disease prevention and control centres (epidemic prevention stations), blood gathering and supplying institutions, health supervision and inspection (check up) institutions, medicinal scientific research and on-job training institutions, health education centres and so on.

10. Medical Organizations include: hospitals, health service centres (stations) in communities, sanatoria, health centres, out-patient clinics, clinics (health stations and infirmaries), maternity and child care agencies (centres and stations), special disease prevention and curing agencies (centres and stations), first aid centres (stations) and clinical inspection centres. Medical organizations are grouped by two types: profit-making and non-profit-making medical organizations.

11. Energy Consumption per Unit of GDP refers to the energy consumption per unit of Gross Domestic Product in a country or the Gross Regional Product in a region in the same reference period. The formula is:

$$\text{Energy Consumption per Unit of GDP} = \frac{\text{Total Energy Consumption}}{\text{Gross Domestic Product}}$$

12. Electricity Consumption per Unit of GDP refers to the electricity consumption per unit of Gross Domestic Product in a country or the Gross Regional Product in a region in the same reference period. The formula is:

$$\text{Electricity Consumption per Unit of GDP} = \frac{\text{Total Electricity Consumption}}{\text{Gross Domestic Product}}$$

13. Energy Consumption per Unit of Industrial Value-added refers to the energy consumption per unit of industrial value-added in a country or region in the same reference period. The formula is:

$$\text{Energy Consumption per Unit of Industrial Value-added} = \frac{\text{Industry Energy Consumption}}{\text{Industrial Value-added.}}$$

14. Gross Amount of Water Used refers to gross water use distributed to users, including loss during transportation, broken down into use by agriculture, industry, living consumption and ecological protection.

15. Total Investment in Fixed Assets in the Whole Country refers to the volume of activities in construction and purchases of fixed assets of the whole country and related fees, expressed in monetary terms during the reference period. It is a comprehensive indicator which shows the size, structure and growth of the investment in fixed assets, providing a basis for observing the progress of construction projects and evaluating results of investment. Total investment in fixed assets in the whole country includes, by type of ownership, the investment by State-owned units, collective-owned units, individuals, joint ownership units, share-holding units, as well as investments by entrepreneurs from foreign countries and from Hong Kong, Macao and Taiwan, and by other units.

16. Total Imports and Exports at Customs refer to the real value of commodities imported and exported across the border of China. They include the actual imports and exports through foreign trade, imported and exported goods under the processing and assembling trades and materials, supplies and gifts as aid given gratis between governments and by the United Nations and other international organizations, and contributions donated by overseas Chinese compatriots in Hong Kong and Macao and Chinese with foreign citizenship, leasing commodities owned by tenant at the expiration of leasing period, the imported and exported commodities processed with imported materials, commodities trading in border areas (excluding mutual exchange goods), the imported and exported commodities and articles for public use of the Sino-foreign joint ventures, cooperative enterprises and ventures with sole foreign investment. Also included is the import or export of samples and advertising goods for which the CIF or FOB value is beyond the permitted ceiling (excluding goods of no trading or use value and free commodities for export), imported goods sold in China from bonded warehouses and other imported or exported goods. The indicator of the total imports and exports at customs can be used to observe the total size of external trade in a country. In accordance with the stipulation of the Chinese government, imports are calculated at CIF, while exports are calculated at FOB.

17. Import-Export Value by Location of China's Foreign Trade Managing Units refers to actual value of imports and exports carried out by corporations which have been registered by the local customs house and are vested with right to run import export business.

18. Disposable Income of Urban Households refers to the actual income at the disposal of members of the households which can be used for final consumption, other non-compulsory expenditure and savings. This equals to total income minus income tax, personal contribution to social security and subsidy for keeping diaries in being a sample household. The following formula is used:

Disposable income = total household income − income tax − personal contribution to social security-subsidy for keeping diaries for a sampled household

19. Consumption Expenditure of Urban Households refers to total expenditure of households for consumption in daily life, including expenditure on the eight categories of food; clothing; household appliances and services; health care and medical services; transport and communications; recreation, education and cultural services; housing; and miscellaneous goods and services.

20. Total Income refers to the sum of income earned from various sources by the rural households and their members during the reference period, and is classified as income from wages and salaries, household operations, properties and transfers.

21. Net Income refers to the total income of rural households from all sources minus all corresponding expenses. The formula for calculation is as follows:

Net income = total income − taxes and fees paid − household operation expenses − taxes and fees − depreciation of fixed assets for production − gifts to non-rural relatives

Net income is mainly used as input for reinvestment in production and as consumption expenditure of the year, and also used for savings and non-compulsory expenses of various forms. "Per capita net income of farmers" is the level of net income averaged by population, reflecting the average income level of rural households in a given area.

22. Total Population refers to the total number of people alive at a certain point of time within a given area.

The annual statistics on total population is taken at midnight, the 3lst of December, not including residents in Taiwan province, Hong Kong and Macao and overseas Chinese.

23. Urban Population refers to all people residing in cities and towns.

24. Employed Persons refers to persons aged 16 and over who are engaged in gainful employment and thus receive remuneration payment or earn business income. This indicator reflects the actual utilization of total labour force during a certain period of time and is often used for the research on China’s economic situation and national power.

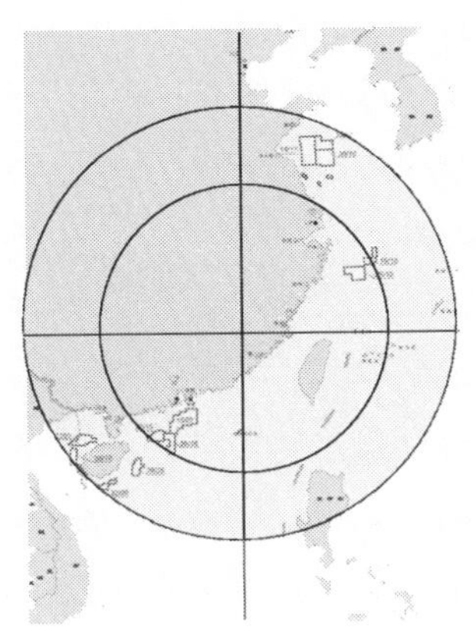

11

部分世界海洋经济统计资料

Part of the World′s Marine Economic Statistics Data

11-1 世界海洋面积
World Ocean Area

区 域 Region	海洋面积 （平方千米） Ocean Area (km^2)	占世界海洋面积的比重（%） Proportion in the World Ocean Area (%)	占地球表面面积的比重（%） Proportion in the Earth's Surface Area (%)
合 计 Total	**361 000 000**	**100.0**	**70.8**
太 平 洋 Pacific Ocean	178 334 000	49.4	35.0
大 西 洋 Atlantic Ocean	91 694 000	25.4	18.0
印 度 洋 Indian Ocean	76 171 000	21.1	14.9
北 冰 洋 Arctic Ocean	14 801 000	4.1	2.9

资料来源：《2011国际统计年鉴》。
Data Source: *International Statistical Yearbook 2011*.

11-2　主要沿海国家（地区）海岸线长度
Length of Coastline of Major Coastal Countries (Regions)

单位：千米　　(km)

国家和地区 Country and Region	海岸线长度 Length of Coastline
中国 China	32 000
美国 United States	22 680
日本 Japan	30 000
德国 Germany	1 300
英国 United Kingdom	11 450
法国 France	3 000
意大利 Italy	7 000
加拿大 Canada	20 000
澳大利亚 Australia	20 125
俄罗斯联邦 Russian Fed.	34 000
波兰 Poland	491
印度 India	6 083
印度尼西亚 Indonesia	35 000
菲律宾 Philippines	18 533
泰国 Thailand	3 219
马来西亚 Malaysia	4 675
新加坡 Singapore	193
缅甸 Myanmar	3 060
孟加拉国 Bangladesh	580
土耳其 Turkey	7 200
韩国 Korea, Rep.	2 413
埃及 Egypt	2 450
墨西哥 Mexico	9 330
巴西 Brazil	7 400
阿根廷 Argentina	4 989

11-3 世界主要沿海国家（地区）国土面积和人口（2012年）
Surface Area and Population of Major Coastal Countries (Regions), 2012

国家和地区 Country and Region	国土面积(万平方千米) Area of Territory (10 000 km^2)	年中人口（万人） Mid-year Population (10 000 persons)	人口密度（人/平方千米） Population Density (people per km^2)
世界总计 World Total	**13 429.2**	**704 390**	**54**
中国 China	960.0	135 070	145
文莱 Brunei Darsm	0.6	41	78
柬埔寨 Cambodia	18.1	1 486	84
印度 India	328.7	123 669	416
印度尼西亚 Indonesia	190.5	24 686	136
日本 Japan	37.8	12 756	350
韩国 Korea, Rep.	10.0	5 000	515
马来西亚 Malaysia	33.1	2 924	89
缅甸 Myanmar	67.7	5 280	81
菲律宾 Philippines	30.0	9 671	324
新加坡 Singapore	0.1	531	7 589
泰国 Thailand	51.3	6 679	131
越南 Viet Nam	33.1	8 878	286
埃及 Egypt	100.2	8 072	81
南非 South Africa	121.9	5 119	43
加拿大 Canada	998.5	3 448	4
墨西哥 Mexico	196.4	12 085	62
美国 United States	983.2	31 391	34
阿根廷 Argentina	278.0	4 109	15
巴西 Brazil	851.5	19 866	23
法国 France	54.9	6 570	120
德国 Germany	35.7	8 189	231
意大利 Italy	30.1	6 092	202
俄罗斯联邦 Russian Fed.	1 709.8	14 353	9
西班牙 Spain	50.6	4 622	94
土耳其 Turkey	78.4	7 400	96
乌克兰 Ukraine	60.4	4 559	79
英国 United Kingdom	24.4	6 323	263
澳大利亚 Australia	774.1	2 268	3
新西兰 New Zealand	26.8	443	17

资料来源：《2014年中国统计年鉴》。

Data Source: *China Statistical Yearbook 2014*.

11-4 主要沿海国家（地区）国内生产总值
Gross Domestic Product of Major Coastal Countries (Regions)

国家和地区 Country and Region	国内生产总值 （亿美元） GDP (100 million USD)	人均国内生产总值 （美元） GNI per Captia (current USD)	国内生产总值增长率 (%) Growth Rate of GDP(%)
中国 China	92 403	6 807	7.67
中国香港 Hong Kong,China	2 740	38 124	2.93
文莱 Brunei Darsm	161	38 563	-1.75
柬埔寨 Cambodia	152	1 008	7.46
印度 India	18 768	1 499	5.02
印度尼西亚 Indonesia	8 683	3 475	5.78
日本 Japan	49 015	38 492	1.54
韩国 Korea, Rep.	13 046	25 977	2.97
马来西亚 Malaysia	3 124	10 514	4.69
菲律宾 Philippines	2 720	2 765	7.16
新加坡 Singapore	2 979	55 182	3.85
泰国 Thailand	3 873	5 779	1.77
越南 Viet Nam	1 714	1 911	5.42
埃及 Egypt	2 720	3 314	2.10
南非 South Africa	3 506	6 618	1.89
加拿大 Canada	18 251	51 911	2.01
墨西哥 Mexico	12 609	10 307	1.07
美国 United States	168 000	53 143	1.88
阿根廷 Argentina	6 118	14 760	2.95
巴西 Brazil	22 457	11 208	2.49
法国 France	27 349	41 421	0.21
德国 Germany	36 348	45 085	0.43
意大利 Italy	20 713	34 619	-1.85
荷兰 Netherlands	8 002	47 617	-0.82
波兰 Poland	5 175	13 432	1.57
俄罗斯联邦 Russian Fed.	20 968	14 612	1.32
西班牙 Spain	13 583	29 118	-1.22
土耳其 Turkey	8 202	10 946	4.05
乌克兰 Ukraine	1 774	3 900	1.88
英国 United Kingdom	25 223	39 351	1.66
澳大利亚 Australia	15 606	67 468	2.66
新西兰 New Zealand	1 826	40 842	2.51

资料来源：世界银行WDI数据库。

Data Source: *World Bank WDI Database.*

11-5 主要沿海国家（地区）国内生产总值产业构成
Industrial Composition of GDP of Major Coastal Countries (Regions) by Industry

国家和地区 Country and Region	国内生产总值产业构成（%） Industrial Structure of GDP(%)		
	第一产业 Primary Industy	第二产业 Secondary Industry	第三产业 Tertiary Industy
世界 World	**3.1**①	**26.7**①	**70.2**①
中国 China	10.0	43.9	46.1
中国香港 Hong Kong,China	0.1②	7.1②	92.9②
文莱 Brunei Darsm	0.7	68.2	31.0
柬埔寨 Cambodia	35.6②	24.3②	40.2②
印度 India	18.2	24.8	57.0
印度尼西亚 Indonesia	14.4	45.7	39.9
日本 Japan	1.2②	25.6②	73.1②
韩国 Korea, Rep.	2.3	38.6	59.1
马来西亚 Malaysia	9.3	40.6	50.1
缅甸 Myanmar	36.4③	26.0③	37.6③
菲律宾 Philippines	11.8②	31.1②	57.1②
新加坡 Singapore		25.1	74.9
斯里兰卡 Sri Lanka	10.8	32.5	56.8
泰国 Thailand	12.0	42.5	45.5
越南 Viet Nam	18.4	38.3	43.3
埃及 Egypt	14.5	39.2	46.3
南非 South Africa	2.4	27.6	70.0
加拿大 Canada			
墨西哥 Mexico	3.5	34.8	61.7
美国 United States	1.2①	20.2①	78.6①
阿根廷 Argentina	6.7	28.6	64.7
巴西 Brazil	5.7	25.0	69.3
法国 France	1.8	18.8	79.4
德国 Germany	0.8	30.2	69.0
意大利 Italy	2.1	23.9	74.0
荷兰 Netherlands	1.6	24.4	74.0
俄罗斯联邦 Russian Fed.	3.9	36.2	59.9
西班牙 Spain	2.6	25.3	72.1
土耳其 Turkey	8.5	27.1	64.4
英国 United Kingdom	0.6	20.3	79.0
澳大利亚 Australia	2.4	27.1	70.5
新西兰 New Zealand	7.2③	23.8③	69.1③

注：①2011年数据。②2012年数据。③2010年数据。

资料来源：世界银行WDI数据库。

Note: ①Data for 2011. ②Data for 2012. ③Data for 2010.

Data Source: *World Bank WDI Database.*

11-6　主要沿海国家（地区）就业人数
Employment in the Major Coastal Countries (Regions)

单位：万人 (10 000 persons)

国家和地区 Country and Region	2000	2005	2007	2008	2009	2010	2011	2012
中国 China	72 085	74 647	76 990	75 564	75 828	76 105	76 420	76 704
中国香港 Hong Kong,China	321	334	319	353	347	347	358	366
印度 India	33 020	37 199				37 429		
印度尼西亚 Indonesia	8 984	9 396	9 993	10 255	10 487	10 821	10 967	
以色列 Israel	222	249	268	278	284	294	302	336
日本 Japan	6 446	6 356	6 412	6 385	6 282	6 257	6 289	6 270
韩国 Korea, Rep.	2 116	2 286	2 343	2 358	2 351	2 383	2 424	2 468
马来西亚 Malaysia	932	1 005	1 102	1 066	1 090	1 178	1 228	1 272
菲律宾 Philippines	2 745	3 231	3 356	3 409	3 506	3 604	3 719	
新加坡 Singapore	209	165	180	185	187		200	204
斯里兰卡 Sri Lanka	631	752	704	765	760	771	820	813
泰国 Thailand	3 300	3 630	3 625	3 784	3 771	3 804	3 932	3 958
越南 Viet Nam	3 837	4 253						
埃及 Egypt	1 720	1 934	2 153	2 251	2 298	2 383	2 335	2 360
南非 South Africa	1 224	1 230	1 347	1 371	1 346	1 306	1 326	1 352
加拿大 Canada	1 476	1 617	1 681	1 713	1 681	1 704	1 731	1 751
墨西哥 Mexico	3 804	4 079	4 306	4 387	4 291	4 387	4 689	4 900
美国 United States	13 521	14 173	14 605	14 536	13 988	13 906		
阿根廷 Argentina	826	964	1 016	1 028	1 034	1 516	1 531	1 570
巴西 Brazil	6 563	8 719	2 044	9 239	9 269	2 202	9 349	9 471
委内瑞拉 Venezuela	896	1 073	1 132	1 186	1 194	1 207	1 239	1 257
法国 France	2 312	2 495	2 558	2 590	2 564	2 569	2 578	2 580
德国 Germany	3 632	3 636	3 979	3 854	3 847	3 874	3 974	4 006
意大利 Italy	2 093	2 256	2 322	2 340	2 303	2 287	2 297	2 290
荷兰 Netherlands	786	811	731	859	860	837	837	842
波兰 Poland	1 452	1 412	1 524	1 580	1 587	1 596	1 613	1 559
俄罗斯联邦 Russian Fed.	6 507	6 817	7 055	7 097	6 928	6 980	7 086	7 155
西班牙 Spain	1 544	1 897	2 036	2 026	1 889	1 846	1 810	1 728
土耳其 Turkey	2 158	2 205	2 079	2 119	2 128	2 259	2 410	2 482
乌克兰 Ukraine	2 018	2 068	2 091	2 097	2 019	2 027	2 032	2 035
英国 United Kingdom	2 726	2 867	2 923	2 936	2 892	2 894	2 908	2 943
澳大利亚 Australia	895	997	1 058	1 074	1 092	1 121	1 136	1 148
新西兰 New Zealand	178	208	217	219	216	218	222	222

资料来源：联合国ILO数据库。

Data Source: *ILO Database.*

11-7 主要沿海国家（地区）鱼类产量（2012年）
Fish Yields of Major Coastal Countries (Regions), 2012

单位：万吨 (10 000 tons)

国家和地区 Country and Region	鱼类产量 Output of Total Fishes	海域鱼类产量 Ocean Area
中国 China	3 580.6	1 082.9
印度 India	815.5	292.9
印度尼西亚 Indonesia	786.3	539.7
缅甸 Myanmar	435.3	228.4
俄罗斯联邦 Russian Fed.	432.0	392.6
美国 United States	422.5	402.2
越南 Viet Nam	413.0	184.7
日本 Japan	316.8	311.3
孟加拉国 Bangladesh	300.9	58.4
菲律宾 Philippines	279.0	235.7
泰国 Thailand	189.4	129.2
马来西亚 Malaysia	142.6	127.0
巴西 Brazil	137.2	50.1
墨西哥 Mexico	136.3	122.3
埃及 Egypt	134.7	9.7
韩国 Korea, Rep.	126.9	124.6
西班牙 Spain	90.6	88.5
尼日利亚 Nigeria	88.5	31.9
南非 South Africa	69.4	69.1
英国 United Kingdom	64.4	63.1
柬埔寨 Cambodia	61.2	9.3
土耳其 Turkey	56.2	41.7
阿根廷 Argentina	52.4	50.7
加拿大 Canada	52.0	48.4
斯里兰卡 Sri Lanka	45.7	38.3
新西兰 New Zealand	41.0	40.7
法国 France	38.6	34.5
荷兰 Netherlands	32.9	32.3

资料来源：联合国FAO数据库。
Data Source: *FAO Database.*

11-8 国家保护区面积和鱼类濒危物种（2013年）
Area of National Nature Reserves and Endangered Species of Fish (2013)

国家和地区 Country and Region	国家保护区① National Protected①		鱼类濒危物种（种） Endangered Species of Fish (number)
	陆地保护区面积占陆地面积比重 Terrestral Protected Areas (% of Total Land Area)	海洋保护区面积占领海面积比重 Marine Protected Areas (% of Territorial Waters)	
中国 China	16.7		121
中国香港 Hong Kong, China	41.9		13
孟加拉国 Bangladesh	4.7	0.1	18
文莱 Brunei Darsm	44.0	15.7	7
柬埔寨 Cambodia	26.2	6.5	40
印度 India	5.2	5.8	213
印度尼西亚 Indonesia	14.7	2.2	145
伊朗 Iran	7.2		31
日本 Japan	16.5	30.0	66
韩国 Korea, Rep.	6.3	0.2	18
马来西亚 Malaysia	18.4		71
菲律宾 Philippines	10.9	52.8	72
新加坡 Singapore	5.4		25
斯里兰卡 Sri Lanka	22.0		43
泰国 Thailand	18.8		96
越南 Viet Nam	6.5		73
埃及 Egypt	11.2	9.5	40
尼日利亚 Nigeria	14.1	1.2	60
南非 South Africa	6.2	9.4	87
加拿大 Canada	8.6		36
墨西哥 Mexico	12.9	0.1	154
美国 United States	13.8	2.0	236
阿根廷 Argentina	6.9		37
巴西 Brazil	26.3	2.0	84
委内瑞拉 Venezuela	53.0	1.7	37
法国 France	24.7	12.1	46
德国 Germany	48.0	1.7	23
意大利 Italy	21.6	4.6	47
荷兰 Netherlands	19.6	0.5	13
波兰 Poland	34.2	4.1	7
俄罗斯联邦 Russian Fed.	11.3		36
西班牙 Spain	29.0	1.3	70
土耳其 Turkey	2.1		70
乌克兰 Ukraine	4.1	9.2	21
英国 United Kingdom	27.9	18.2	43
澳大利亚 Australia	12.9		106
新西兰 New Zealand	27.3	37.7	23

注：①2012年数据。
Note: ①Data refer to 2012.
资料来源：世界银行WDI数据库。
Data Source: *World Bank WDI Database..*

11-9 主要沿海国家（地区）风力发电量
Wind-Power Capacifies of Major Coastal Countries (Regions)

单位：百万千瓦·时 (million kilowatt hours)

国家和地区 Country and Region	风电 wind				
	2006	2007	2008	2009	2010
中国 China			13 079	46 096	44 622
伊朗 Iran		143	196	224	
以色列 Israel	2	1	9	9	8
日本 Japan	1 753	2 624	2 623	2 949	3 962
韩国 Korea, Rep	239	376	436	685	817
菲律宾 Philippines	55	59	61	64	62
斯里兰卡 Sri Lanka	2	2	3	3	53
泰国 Thailand	1				3
埃及 Egypt	616	831	931	1 133	1 498
尼日利亚 Nigeria					
南非 South Africa	32	32	32	32	32
加拿大 Canada	2 500	3 024	3 819	4 573	9 557
墨西哥 Mexico	59	262	269	596	1 239
美国 United States	26 676	34 603	55 696	74 226	95 148
阿根廷 Argentina	70	61	42	36	25
法国 France	2 150	4 052	5 689	7 891	9 969
德国 Germany	30 710	39 713	40 574	38 639	37 793
意大利 Italy	2 971	4 034	4 861	6 543	9 126
荷兰 Netherlands	2 733	3 438	4 260	4 581	3 993
波兰 Poland	256	522	837	1 077	1 664
俄罗斯联邦 Russian Fed.	5	7	5	4	4
西班牙 Spain	23 040	27 509	32 203	37 773	44 165
土耳其 Turkey	127	355	847	1 495	2 916
乌克兰 Ukraine	35	45	45	43	50
英国 United Kingdom	4 225	5 274	7 097	9 304	10 183
澳大利亚 Australia	1 691	2 611	3 941	3 806	4 798
新西兰 New Zealand	623	937	1 057	1 471	1 634

资料来源：联合国ESD数据库。
Data Source: UN ESD Database.

11-10 主要沿海国家（地区）捕捞产量
Fishing Yields in the Major Coastal Countries(Regions)

单位：吨 (t)

国家和地区 Country and Region	2011	2012
世界总计 World Total	**93 734 327**	**91 336 230**
中国 China	15 768 630	16 167 443
秘鲁 Peru	8 248 482	4 841 524
印度尼西亚 Indonesia	5 701 440	5 813 800
美国 United States	5 153 452	5 128 381
印度 India	4 311 132	4 862 861
俄罗斯联邦 Russian Fed.	4 254 877	4 331 398
日本 Japan	3 775 545	3 644 328 *
缅甸 Myanmar	3 332 979	3 579 250
智利 Chile	3 063 467	2 572 881
越南 Viet Nam	2 514 300	2 622 200
菲律宾 Philippines	2 363 228	2 322 850
挪威 Norway	2 282 608	2 150 555
泰国 Thailand	1 835 126	1 834 573
韩国 Korea, Rep.	1 748 153	1 670 385
孟加拉国 Bangladesh	1 600 918	1 535 715
墨西哥 Mexico	1 566 063	1 575 409
马来西亚 Malaysia	1 378 799	1 477 281
冰岛 Iceland	1 138 462	1 449 587
西班牙 Spain	1 004 965	930 018
摩洛哥 Morocco	958 907	1 171 496
中国台湾 Taiwan China	903 920	907 638
加拿大 Canada	865 286	814 946
巴西 Brazil	803 267	842 987
阿根廷 Argentina	793 308	738 060
南非 South Africa	533 432	701 711
尼日利亚 Nigeria	635 486	668 754
英国 United Kingdom	600 536	631 442
柬埔寨 Cambodia	560 839	566 695
伊朗 Iran	487 817	542 378
丹麦 Denmark	716 312	502 729
斯里兰卡 Sri Lanka	426 768	473 832
新西兰 New Zealand	429 836	440 683
土耳其 Turkey	514 763	432 444
法国 France	448 221	425 694
埃及 Egypt	375 354	354 237
荷兰 Netherlands	370 097	347 344
委内瑞拉 Venezuela	201 785	213 072
德国 Germany	233 883	207 500
意大利 Italy	216 939	200 816

注：捕捞品种包括鱼类、甲壳类、软体类等水生动物。*为联合国粮农组织估算值。

资料来源：《渔业和水产养殖统计年鉴》，联合国粮农组织，2012年(表11-11同)。

Note: The fished species include fish, crustacea, mollusc and aquatic animals.

* It is estimated by FAO from available sources of information or calculation.

Source: *Fishery and AquacultureStatistical Yearbook*, FAO, 2012 (Same as Table 11-11).

表11-11 主要沿海国家水产养殖产量
Aquaculture Production in the Major Coastal Countries(Regions)

单位：吨 (t)

国家和地区 Country and Region	2011	2012
世界总计 World Total	**62 011 524**	**66 633 253**
中国 China	38 621 269	41 108 306
印度 India	3 673 082	4 209 415
越南 Viet Nam	2 845 600	3 085 500
印度尼西亚 Indonesia	2 718 421	3 067 660
孟加拉国 Bangladesh	1 523 759	1 726 066
挪威 Norway	1 143 893	1 321 119
泰国 Thailand	1 201 455	1 233 877
智利 Chile	954 845	1 071 421
埃及 Egypt	986 820	1 017 738
缅甸 Myanmar	816 820	885 169
菲律宾 Philippines	767 287	790 894
巴西 Brazil	629 609	707 461
日本 Japan	556 761	633 047
韩国 Korea, Rep.	507 052	484 404
美国 United States	397 292	420 024
中国台湾 Taiwan China	314 363	344 404
伊朗 Iran	247 262	296 575
马来西亚 Malaysia	287 276	283 780
西班牙 Spain	271 961	264 160
尼日利亚 Nigeria	221 128	253 898
土耳其 Turkey	188 890	212 805
法国 France	206 875	204 860
英国 United Kingdom	198 439	203 037
加拿大 Canada	162 414	173 452
意大利 Italy	164 151	162 618
俄罗斯联邦 Russian Fed.	128 830	144 871
墨西哥 Mexico	137 130	143 747

注：养殖品种包括鱼类、甲壳类、软体类等水生动物。

Note: The cultivated varieties includes fish, crustacea, mollusc and other aquatic animals.

11-12 主要沿海国家（地区）石油主要指标（2012年）
Main Oil Indicators of Major Coastal Countries (Regions), 2012

国家和地区 Country and Region	原油探明储量（亿桶） Crude Oil Proved Reserves (100 million Barrels)	石油存量（万桶） Total Petroleum Stocks (10 000 Barrels)
世界 World	**15 260**	**418 845**
中国 China	204	
孟加拉国 Bangladesh		
文莱 Brunei Darsm		
印度 India	56	
印度尼西亚 Indonesia	39	
伊朗 Iran	1 512	
以色列 Israel		
日本 Japan		59 065
韩国 Korea, Rep.		17 544
马来西亚 Malaysia	40	
缅甸 Myanmar		
巴基斯坦 Pakistan		
菲律宾 Philippines		
泰国 Thailand		
越南 Viet Nam	44	
埃及 Egypt	44	
尼日利亚 Nigeria	372	
南非 South Africa		
加拿大 Canada	1 736	17 410
墨西哥 Mexico	104	4 750
美国 United States	265	180 778
阿根廷 Argentina	25	
巴西 Brazil	140	
委内瑞拉 Venezuela		
法国 France		16 234
德国 Germany		28 712
意大利 Italy		12 550
荷兰 Netherlands		12 132
波兰 Poland		6 390
俄罗斯联邦 Russian Fed.	600	
西班牙 Spain		12 007
土耳其 Turkey		6 203
乌克兰 Ukraine		
英国 United Kingdom	28	8 086
澳大利亚 Australia		3 754
新西兰 New Zealand		784

资料来源：美国能源署。

Data Source: U. S.Energy Information Administration.

11-13 主要沿海国家（地区）能源利用效率
Energy Efficiency in the Major Coastal Countries (Regions)

单位：吨标准油/万美元 (TOE per 10 000 USD)

国家和地区 Country and Region	2005年不变价 (constant 2005 US$)		
	2000	2010	2011
世界 World	**2.46**	**2.45**	**2.42**
中国 China	8.20	6.56	6.50
中国香港 Hong Kong,China	0.91	0.63	0.65
孟加拉国 Bangladesh	4.02	3.78	3.60
文莱 Brunei Darsm	2.77	3.29	3.81
柬埔寨 Cambodia	8.47	5.78	5.73
印度 India	7.60	5.80	5.65
印度尼西亚 Indonesia	6.82	5.59	5.19
伊朗 Iran	8.41		
以色列 Israel	1.51	1.40	1.34
日本 Japan	1.20	1.07	1.00
韩国 Korea, Rep.	2.77	2.45	2.46
马来西亚 Malaysia	4.14	4.08	4.05
巴基斯坦 Pakistan	7.46	6.29	6.15
菲律宾 Philippines	4.84	3.09	2.98
新加坡 Singapore	1.91	2.02	1.88
斯里兰卡 Sri Lanka	4.14	2.96	2.90
泰国 Thailand	5.26	5.59	5.67
越南 Viet Nam	7.80	7.93	7.78
埃及 Egypt	5.39	6.08	6.30
尼日利亚 Nigeria	10.87	7.41	7.10
南非 South Africa	5.34	4.91	4.72
加拿大 Canada	2.51	2.08	2.04
墨西哥 Mexico	1.88	1.94	1.94
美国 United States	2.04	1.71	1.66
阿根廷 Argentina	3.67		
巴西 Brazil	2.44	2.42	2.40
委内瑞拉 Venezuela	4.40	4.33	3.86
法国 France	1.28	1.18	1.12
德国 Germany	1.25	1.11	1.02
意大利 Italy	1.01	0.97	0.95
荷兰 Netherlands	1.22	1.22	1.12
波兰 Poland	3.41	2.65	2.53
俄罗斯联邦 Russian Fed.	10.91	7.72	7.71
西班牙 Spain	1.27	1.08	1.06
土耳其 Turkey	1.97	1.86	1.83
乌克兰 Ukraine	22.47	14.61	13.28
英国 United Kingdom	1.12	0.85	0.79
澳大利亚 Australia	1.83	1.54	1.50
新西兰 New Zealand	1.82	1.55	1.52

资料来源：世界银行WDI数据库。

Data Source: *World Bank WDI Database.*

11-14 主要沿海国家（地区）国际海运装货量和卸货量
International Maritime Freight Loaded and Unloaded in the Major Coastal Countries (Regions)

单位：万吨 (10 000 t)

国家和地区 Country and Region	国际海运装货量 International Ocean Shipping Loading Capacity			国际海运卸货量 International Ocean Shipping Unloading Capacity		
	2000	2005	2013	2000	2005	2013
中国香港 Hong Kong,China	6 770	8 918	11 255	10 693	14 095	16 020
孟加拉国 Bangladesh	89	79	466 ①	1 408		3 860 ①
文莱 Brunei Darsm	10	8		102	168	
印度尼西亚 Indonesia	14 153	27 372	50 118 ①	4 504	8 479	11 254 ①
伊朗 Iran	3 065	3 446		4 486	6 073	
以色列 Israel	1 387	1 663	2 046	2 920	2 107	2 828
日本 Japan	13 010			80 654		
韩国 Korea Rep.	15 078	24 250		41 882	51 245	
马来西亚 Malaysia	5 483	7 940	13 253	6 922	10 391	15 770
巴基斯坦 Pakistan	617	1 126	1 950 ①	3 080	3 955	4 870 ①
新加坡 Singapore	32 618	42 266				
斯里兰卡 Sri Lanka	919	1 315				
埃及 Egypt		2 123			4 241	
南非 South Africa	2 598					
美国 United States	34 334	40 534		83 352	94 160	
阿根廷 Argentina	1 550					2 747
法国 France	6 810	10 066	9 989	20 273	22 710	19 040
德国 Germany	8 602	10 832	11 860	14 725	16 866	17 108
荷兰 Netherlands	9 940	12 250		32 507	36 424	
波兰 Poland	3 152	3 835	3 025	1 582	1 642	3 311
俄罗斯联邦 Russian Fed.	828	910	18 924 ②	84	74	2 184 ②
西班牙 Spain	5 627	8 195		19 343	26 164	
乌克兰 Ukraine	4 271	7 070	9 920	684	1 333	1 862
澳大利亚 Australia	48 750	62 401	113 288	5 418	6 989	9 740
新西兰 New Zealand	2 214	2 167	3 656	1 379	1 844	2 006

注：①2010年数据。②2011年数据。
资料来源：联合国统计月报数据库。

Note: ①Data for 2010. ②Data for 2011.
Data Source: *UN Monthly Bulletin of Statistics Database.*

11-15 世界主要外贸货物海运量及构成*（2012年）
World Major Maritime Freight Traffic in Foreign Trade*, 2012

品种 Sort	海运量（百万吨） Freight Traffic (million tons)	
	2012	所占比例（%） Percentage
合　计 **Total**	**9 499**	**100.0**
原　油 Crude Oil	1 856	19.5
成品油 Refined Oil	887	9.3
燃　气 Fuel gas	304	3.2
铁矿石 Ironstone	1 115	11.7
煤　炭 Coal	1 031	10.9
谷　物 Corn	357	3.8
其他货物 Others	3 949	41.6

注：*为估计数。
资料来源：Shipping Statistics and Market Review, January/February 2013, ISL
Note: *Estimated data.
Data Source: *Shipping Statistics and Market Review*, January/February 2013, ISL.

11-16 主要沿海国家（地区）国际旅游人数
Number of International Tourists of Major Coastal Countries (Regions)

单位：万人 (10 000 persons)

国家和地区 Country and Region	入境（过夜）旅游人数 Number of Inbound Tourists (Overnight)			出境旅游人数 Number of Outbound Tourists		
	2005	2010	2012	2005	2010	2012
世界总计 World Total	**80 493**	**98 494**	**107 662**	**85 387**	**109 620**	**120 307**
中国 China	4 681	5 566	5 773	3 103	5 739	8 318
中国香港 Hong Kong, China	1 477	2 009	2 377	7 230	8 444	8 528
中国澳门 Macao, China	901	1 193	1 358	30	75	129
孟加拉国 Bangladesh	21	30		177		
文莱 Brunei Darsm	13	21	21			
柬埔寨 Cambodia	133	251	358	57	51	79
印度 India	392	578	658	719	1 299	1 492
印度尼西亚 Indonesia	500	700	804	411	624	745
伊朗 Iran	189	294				
以色列 Israel	190	280	289	369	427	435
日本 Japan	673	861	836	1 740	1 664	1 849
韩国 Korea, Rep.	602	880	1 114	1 008	1 249	1 374
马来西亚 Malaysia	1 643	2 458	2 503			
缅甸 Myanmar	23	31	59			
巴基斯坦 Pakistan	80	91	97			
菲律宾 Philippines	262	352	427	214		
新加坡 Singapore	708	916	1 110	516	734	805
斯里兰卡 Sri Lanka	55	65	101	73	112	127

11-16 续表 continued

国家和地区 Country and Region	入境（过夜）旅游人数 Number of Inbound Tourists (Overnight)			出境旅游人数 Number of Outbound Tourists		
	2005	2010	2012	2005	2010	2012
泰国 Thailand	1 157	1 594	2 235	305	545	572
越南 Viet Nam	348	505	685			
埃及 Egypt	824	1 405	1 120	531	462	414
尼日利亚 Nigeria	101	611	467			
南非 South Africa	737	807	919		517	503
加拿大 Canada	1 877	1 622	1 634	2 110	2 868	3 228
墨西哥 Mexico	2 192	2 329	2 340	1 331	1 433	1 558
美国 United States	4 921	5 980	6 697	6 350	6 027	6 072
阿根廷 Argentina	382	533	559	389	531	725
巴西 Brazil	536	516	568	347	645	812
委内瑞拉 Venezuela	71	53	71	107	148	202
法国 France	7 499	7 765	8 301	2 248	2 504	2 545
德国 Germany	2 150	2 688	3 041	7 740		
意大利 Italy	3 651	4 363	4 636	2 480	2 982	2 881
荷兰 Netherlands	1 001	1 088	1 168	1 704	1 837	1 863
波兰 Poland	1 520	1 247	1 484	4 084	4 276	
俄罗斯联邦 Russian Fed.	2 220	2 228	2 818	2 842	3 932	
西班牙 Spain	5 591	5 268	5 770	1 046	1 238	1 219
土耳其 Turkey	2 027	3 136	3 570	825	656	580
乌克兰 Ukraine	1 763	2 120	2 301	1 645	1 718	2 143
英国 United Kingdom	2 804	2 830	2 928	6 649	5 556	5 654
澳大利亚 Australia	550	589	615	476	711	
新西兰 New Zealand	235	244	247	187	203	217

资料来源：世界银行WDI数据库。

Data source: *WDI database of World Bank.*

11-17 主要沿海国家（地区）国际旅游收入
International Tourism Receipts of Major Coastal Countries (Regions)

单位：亿美元 (10 000 million USD)

国家和地区 Country and Region	国际旅游收入 International Tourism Receipts		
	2010	2011	2012
世界总计 World Total	**11 121**	**12 495**	**12 972**
中国 China	502	533	549
中国香港 Hong Kong, China	272	337	380
中国澳门 Macao, China	282	390	445
孟加拉国 Bangladesh	1	1	1
文莱 Brunei Darsm			
柬埔寨 Cambodia	13	18	20
印度 India	145	177	183
印度尼西亚 Indonesia	76	90	95
伊朗 Iran	26	26	
以色列 Israel	58	60	62
日本 Japan	154	125	162
韩国 Korea, Rep.	144	175	197
马来西亚 Malaysia	182	196	203
缅甸 Myanmar	1	3	
巴基斯坦 Pakistan	10	11	10
菲律宾 Philippines	32	40	49
新加坡 Singapore	142	181	193
斯里兰卡 Sri Lanka	10	14	18

11-17 续表 continued

国家和地区 Country and Region	国际旅游收入 International Tourism Receipts		
	2010	2011	2012
泰国 Thailand	238	309	377
越南 Viet Nam	45	57	68
埃及 Egypt	136	93	108
尼日利亚 Nigeria	7	7	6
南非 South Africa	103	107	112
加拿大 Canada	184	200	207
墨西哥 Mexico	126	125	133
美国 United States	1 646	1 845	2 001
阿根廷 Argentina	56	61	57
巴西 Brazil	62	68	69
委内瑞拉 Venezuela	8	8	9
法国 France	561	660	635
德国 Germany	491	534	516
意大利 Italy	401	454	430
荷兰 Netherlands	187	210	205
波兰 Poland	100	116	118
俄罗斯联邦 Russian Fed.	132	170	179
西班牙 Spain	590	677	632
土耳其 Turkey	263	301	323
乌克兰 Ukraine	47	54	60
英国 United Kingdom	408	459	460
澳大利亚 Australia	323	342	341
新西兰 New Zealand	49	55	55

资料来源：世界银行WDI数据库。

Data Source: *World Bank WDI Database..*

11-18 集装箱吞吐量居世界前20位的港口
World Top 20 Seaports in Terms of the Number of Containers Handled

单位：万标准箱 (10 000 TEU)

港 口 Seaport	所属国家或地区 Country or Region	吞吐量 Containers Handled
上海 Shanghai	中国 China	3 362
新加坡 Singapore	新加坡 Singapore	3 258
深圳 Shenzhen	中国 China	2 328
香港 Hong Kong	中国 China	2 235
釜山 Pusan	韩国 Korea, Rep.	1 769
宁波-舟山 Ningbo and Zhoushan	中国 China	1 735
青岛 Qingdao	中国 China	1 552
广州 Guangzhou	中国 China	1 531
迪拜 Dubayy	阿联酋 United Arab Em	1 364
天津 Tianjin	中国 China	1 301
鹿特丹 Rotterdam	荷兰 Netherlands	1 166
巴生 Kelang	马来西亚 Malaysia	1 035
大连 Dalian	中国 China	1 001
高雄 Gaoxiong	中国台湾 Taiwan, China	994
汉堡 Hamburg	德国 Germany	928
安特卫普 Antwerp	比利时 Belgium	858
洛杉矶 Los Angeles	美国 United States	808
厦门 Xiamen	中国 China	801
丹戎帕拉帕斯 Tanjung Periuk	马来西亚 Malaysia	763
雅加达 Jakarta	印度尼西亚 Indonesia	621

资料来源：上海航运交易所，CL-ONLINE。
Data Source: *Shanghai Shipping Exchange, CL-ONLINE.*

11-19 港口货物吞吐量居世界前20位的港口（2012年）
World Top 20 Seaports in Terms of the Cargo Handled, 2012

单位：百万吨 (million t)

港 口 Seaport	所属国家或地区 Country or Region	吞吐量 Cargo Handled	备 注 Remark
上海 Shanghai	中国 China	735.6	内外贸货物 Domestic and Foreign Trade Cargo
新加坡 Singapore	新加坡 Singapore	538.0	内外贸货物 Domestic and Foreign Trade Cargo
天津 Tianjin	中国 China	477.0	内外贸货物 Domestic and Foreign Trade Cargo
宁波 Ningbo	中国 China	453.0	内外贸货物 Domestic and Foreign Trade Cargo
鹿特丹 Rotterdam	荷兰 Netherlands	441.5	内外贸货物 Domestic and Foreign Trade Cargo
广州 Guangzhou	中国 China	435.2	内外贸货物 Domestic and Foreign Trade Cargo
青岛 Qingdao	中国 China	406.9	内外贸货物 Domestic and Foreign Trade Cargo
大连 Dalian	中国 China	374.3	内外贸货物 Domestic and Foreign Trade Cargo
釜山 Pusan	韩国 Korea, Rep.	298.7	内外贸货物 Domestic and Foreign Trade Cargo
黑德兰港 Headland Harbour	澳大利亚 Australia	288.4	内外贸货物 Domestic and Foreign Trade Cargo
秦皇岛 Qinhuangdao	中国 China	271.0	内外贸货物 Domestic and Foreign Trade Cargo
香港 Hong Kong	中国 China	269.3	内外贸货物 Domestic and Foreign Trade Cargo
南路易斯安娜 Southern Louisiana	美国 United States	253.0	外贸货物 Foreign Trade Cargo
休斯敦 Houston	美国 United States	246.9	内外贸货物 Domestic and Foreign Trade Cargo
深圳 Shenzhen	中国 China	228.1	内外贸货物 Domestic and Foreign Trade Cargo
名古屋 Nagoya	日本 Japan	202.6	内外贸货物 Domestic and Foreign Trade Cargo
巴生 Kelang	马来西亚 Malaysia	195.9	内外贸货物 Domestic and Foreign Trade Cargo
安特卫普 Antwerp	比利时 Belgium	184.1	内外贸货物 Domestic and Foreign Trade Cargo
丹皮尔 Dampier	澳大利亚 Australia	180.4	内外贸货物 Domestic and Foreign Trade Cargo
洛杉矶 Los Angeles	美国 United States	175.2	内外贸货物 Domestic and Foreign Trade Cargo

资料来源：Shipping Statistics and Market Review, December 2013, ISL.

Data Source: *Shipping Statistics and Market Review, December 2013, ISL.*

11-20 海上商船拥有量居世界前20位的国家或地区（2012年）
World Top 20 Countries or Regions in Terms of the Number of Maritime Merchant Ships Owned ,2012

国家和地区 Country and Region	艘数 （艘） Number of Vessels (unit)	载重吨 Deadweight ton	
		万吨 10 000 tons	占世界% Percentage in the World
世界总计 World Total	**48 742**	**153 926.3**	**100.0**
巴拿马 Panama	6 979	34 033.9	22.1
利比里亚 Liberia	2 988	19 421.3	12.6
马绍尔群岛 Marshall Islamds	1 878	13 286.6	8.6
中国香港 Hong Kong, China	2 070	12 896.0	8.4
新加坡 Singapore	1 896	8 817.3	5.7
希腊 Greece	1 098	7 596.3	4.9
马耳他 Malta	1 675	6 819.5	4.4
中国 China	2 380	6 386.7	4.1
巴哈马 Bahamas	1 173	6 307.4	4.1
英国 United Kingdom	958	4 110.9	2.7
塞浦路斯 Cyprus	834	3 124.1	2.0
意大利 Italy	801	2 009.0	1.3
日本 Japan	2 684	1 948.5	1.3
挪威 Norway	828	1 873.8	1.2
德国 Germany	451	1 724.5	1.1
韩国 Korea, Rep.	1 051	1 704.3	1.1
印度 India	604	1 496.6	1.0
安提瓜和巴布亚 Antigua and Papua	1 256	1 430.5	0.9
丹麦 Denmark	450	1 369.8	0.9
印度尼西亚 Indonesia	2 227	1 128.9	0.7

注：1. 表中数据按载重吨排序；

2. 统计范围为300总吨及以上船舶，截至日期均为2013年1月1日；

3. 因统计口径不同，表中数据与其他出版物公布的数据略有差别。

资料来源：Shipping Statistics and Market Review, Jannuang/February 2013. ISL.

Note: 1. The data in the table are arranged in the order of deadweight tons.

2. The ships, ranging over 300 tons and more in gross ton, are counted as of Jan. 1, 2013.

3. Owing to different statistical requirements, the data given in the table may be somewhat different from those issued in other publications.

Data Source: *Shipping statistics and Market Review, Jannuang/February 2013. ISL.*

11-21　集装箱船拥有量居世界前20位的国家或地区（2012年）
World Top 20 Countries or Regions in Terms of the Number of Container Ships Owned, 2012

国家和地区 Country and Region	艘数 （艘） Number of Vessels (unit)	载重吨 Deadweight ton	
		万标准箱 10 000 TEU	占世界% Percentage in the World
世界总计 World Total	**5 091**	**1 622.7**	**100.0**
利比里亚 Liberia	998	370.3	22.8
巴拿马 Panama	727	301.7	18.6
中国香港 Hong Kong, China	320	135.2	8.3
德国 Germany	278	120.6	7.4
新加坡 Singapore	349	101.6	6.3
英国 United Kingdom	200	97.2	6.0
马绍尔群岛 Marshall Islamds	236	68.5	4.2
丹麦 Denmark	101	62.2	3.8
安提瓜和巴布亚 Antigua and Papua	409	56.2	3.5
马耳他 Malta	117	45.2	2.8
中国 China	187	44.7	2.8
塞浦路斯 Cyprus	209	40.4	2.5
美国 United States	66	20.6	1.3
希腊 Greece	35	19.7	1.2
法国 France	27	19.4	1.2
巴哈马 Bahamas	54	14.6	0.9
荷兰 Netherlands	71	11.2	0.7
印度尼西亚 Indonesia	158	9.4	0.6
韩国 Korea, Rep.	87	8.9	0.5
意大利 Italy	22	7.8	0.5

注：1. 表中数据按集装箱位排序；
2. 统计范围为300总吨及以上船舶，截至日期均为2013年1月1日；
3. 因统计口径不同，表中数据与其他出版物公布的数据略有差别。

资料来源：Shipping statistics and Market Review, Jannuang/February 2013. ISL.

Note: 1. The data in the table are arranged in the order of deadweight tons.
2. The ships, ranging over 300 tons and more in gross ton, are counted as of Jan. 1, 2013.
3. Owing to different statistical requirements, the data given in the table may be somewhat different from those issued in other publications.

Data Source: *Shipping statistics and Market Review, Jannuang/February 2013. ISL.*

图书在版编目（CIP）数据

中国海洋统计年鉴．2014/国家海洋局编著．--北京：海洋出版社，2015.3

ISBN 978-7-5027-9119-3

Ⅰ．①中… Ⅱ．①国… Ⅲ．①海洋－统计资料－中国－2014－年鉴 Ⅳ．①P7-66

中国版本图书馆 CIP 数据核字(2015)第 058339 号

责任编辑：王　溪

责任印制：赵麟苏

海洋出版社 出版发行

http: //www.oceanpress.com.cn

北京市海淀区大慧寺路 8 号　　邮编：100081

国家海洋信息中心印刷厂印刷　　新华书店发行所经销

2015 年 3 月第 1 版　　2015 年 3 月第 1 次印刷

开本：787mm×1092mm　1/16　印张：19.75

字数：495 千字

定价：168.00 元

发行部：(010)62147016　　邮购部：(010)68038093　　总编室：(010)62114335